Digital techniques

The Digital Trainer
(with permission of Beal-Davis Electronics Ltd.)

Digital techniques

Barry G. Woollard M Phil, C Eng, MIERE, M Inst MC
Technical Director, Beal-Davis Electronics Limited

McGraw-Hill Book Company (UK) Limited

London · New York · St Louis · San Francisco · Auckland · Bogotá
Guatemala · Hamburg · Johannesburg · Lisbon · Madrid · Mexico · Montreal
New Delhi · Panama · San Juan · São Paulo · Singapore · Sydney · Tokyo
Toronto

Published by
McGraw-HILL Book Company (UK) Limited
MAIDENHEAD · BERKSHIRE · ENGLAND

British Library Cataloguing in Publication Data

Woollard, Barry G.
  Digital techniques.
  1. Digital electronics
  I. Title
  621.3815   TK7868.D5

  ISBN 0–07–084669–3

Library of Congress Cataloging in Publication Data

Woollard, Barry G.
  Digital techniques.

  Includes index.
  1. Digital electronics.   I. Title.
TK7868.D5W63   1983   621.3819′58   82–20909
ISBN 0–07–084669–3

12345 JWA 86543

Printed in Great Britain at the Alden Press, Oxford

Contents

Preface

Developments in the technology of integrated circuits during the last decade have caused major changes to be made in the field of digital electronics and micro-computing. These advances have led to the widespread use of electronic calcula-tors, digital watches, and electronic games in the social and domestic field, and to digital multimeters (DMMs), digital frequency counters, digital panel meters (DPMs), direct digital control of industrial processes, programmable logic control (PLC), and the microcomputer—enabling industrial control applications to be solved at costs which can no longer be ignored.

This text is written to meet all of the objectives of the TEC Units: Digital Techniques 2, Digital Techniques 3A, and Digital Techniques 3B. In addition to the TEC requirements, the text is useful for a wide range of technicians and engineers who are interested in gaining some expertise in the use of digital electronics.

In the field of Digital Techniques, and in Microcomputing, it is *essential* that the normal learning process is complemented by 'hands-on' practical development. This text is prepared so that the diagrams can be interpreted as 'practical circuits'; a Digital Trainer is manufactured by Beal-Davis Electronics Ltd., 18A Friar Street, Worcester. This equipment is designed to provide all the necessary power supplies, binary signals, pulse generators, and monitoring capability required to satisfy all of the objectives in this text.

Many examples are included in the text, each chapter concluding with exercises. Answers to numerical exercises are included in the Appendices.

I sincerely acknowledge my gratitude to all those who have contributed to the realization of this book, in particular to my wife, for her patience and efforts during the preparation and typing of the manuscript.

Barry G. Woollard
Longdon, Staffs.

1 Binary arithmetic

1.1 Numbering systems

Many different numbering systems are in common use. These have evolved out of the need to simplify the human operator's task in reading and writing data (information) and coded instructions to digital electronic control equipment and to computer systems. A *digital code* is a system of symbols that represent data values and make up a special 'language' that permits information to be communicated, stored, and manipulated by digital circuits and computers.

The number of characters (digits, symbols) used in a system is known as its *base*, or *radix*. Other numbers are 'constructed' by giving different 'weights' to the position of the digit relative to the 'decimal' point. The 'weights' of the different positions are given by powers of the radix, which, in general is:

$$R^4 \quad R^3 \quad R^2 \quad R^1 \quad R^0 \quad \cdot \quad R^{-1} \quad R^{-2} \quad R^{-3} \quad R^{-4}$$

(a) *Denary numbering system.* Although the denary system is very widely understood, it is helpful to examine it here to appreciate the 'construction' technique used, so that numbering systems which are unfamiliar can be better understood.

The radix is 10, i.e., *ten symbols* are used to represent the quantities 0 (zero) through 9 (nine). Positional weights are powers of 10:

$$10^4 \quad 10^3 \quad 10^2 \quad 10^1 \quad 10^0 \quad \cdot \quad 10^{-1} \quad 10^{-2}$$

EXAMPLE 1.1

$$
\begin{aligned}
2345.63_{10} &= 2 \times 10^3 + 3 \times 10^2 + 4 \times 10^1 + 5 \times 10^0 + 6 \times 10^{-1} + 3 \times 10^{-2} \\
&= 2 \times 1000 + 3 \times 100 + 4 \times 10 + 5 \times 1 + 6 \times 0.1 + 3 \times 0.01 \\
&= 2000 + 300 + 40 + 5 + 0.6 + 0.03 \\
&= 2345.63_{10}
\end{aligned}
$$

(b) *Binary numbering system.* This system is very widely used in logic and computing, and is the basis of machine code language used with computers.

The radix is 2, i.e., *two symbols* are used to represent the quantities 0 and 1. Positional weights are powers of 2:

$$2^7 \quad 2^6 \quad 2^5 \quad 2^4 \quad 2^3 \quad 2^2 \quad 2^1 \quad 2^0 \quad \cdot \quad 2^{-1} \quad 2^{-2} \quad 2^{-3} \quad 2^{-4}$$

EXAMPLE 1.2

$$11101011 \cdot 101_2 = 1 \times 2^7 + 1 \times 2^6 + 1 \times 2^5 + 1 \times 2^3 + 1 \times 2^1 + 1 \times 2^0$$
$$+ 1 \times 2^{-1} + 1 \times 2^{-3}$$
$$= 128 + 64 + 32 + 8 + 2 + 1 + 0.5 + 0.125$$
$$= 235.625_{10}$$

(c) *Octal numbering system.* This system has been widely used in computer systems, due to the ease with which it can be converted to binary and vice versa. Octal was used in preference to binary, since a given number can be represented by fewer digits than are required in binary.

The radix is 8, i.e., *eight symbols* are used to represent the quantities 0 through 7. Positional weights are powers of 8:

$$8^4 \quad 8^3 \quad 8^2 \quad 8^1 \quad 8^0 \quad \cdot \quad 8^{-1} \quad 8^{-2} \quad 8^{-3}$$

EXAMPLE 1.3

$$2345.63_8 = 2 \times 8^3 + 3 \times 8^2 + 4 \times 8^1 + 5 \times 8^0 + 6 \times 8^{-1} + 3 \times 8^{-2}$$
$$= 2 \times 512 + 3 \times 64 + 4 \times 8 + 5 \times 1 + 6 \times 0.125 + 3 \times 0.015625$$
$$= 1024 + 192 + 32 + 5 + 0.75 + 0.046875$$
$$= 1253.796875_{10}$$

(d) *Hexadecimal (hex) numbering system.* This system has become much more widely used than octal, due to the advances made in the development of microcomputer systems. *Hex* has replaced the use of octal in most microcomputers, due to the reduced number of digits necessary to represent any particular binary number.

The radix is 16, i.e., *sixteen symbols* are used to represent the quantities 0, 1, 2, 3, 4, 5, 6, 7, 8, 9, A, B, C, D, E, F:

Decimal	Hexadecimal
0	0
1	1
2	2
3	3
4	4
5	5
6	6
7	7
8	8
9	9
10	A

11	B
12	C
13	D
14	E
15	F

In general:

$$16^4 \quad 16^3 \quad 16^2 \quad 16^1 \quad 16^0 \quad \cdot \quad 16^{-1} \quad 16^{-2} \quad 16^{-3}$$

EXAMPLE 1.4

$$B6EF \cdot 4AH = B \times 16^3 + 6 \times 16^2 + E \times 16^1 + F \times 16^0 + 4 \times 16^{-1} + A \times 16^{-2}$$

$$= 11 \times 4096 + 6 \times 256 + 4 \times 16 + 15 \times 1 + 4 \times 0.0625$$

$$+ 10 \times 0.00390625$$

Hex

radix $16_{10} = 45056 + 1536 + 224 + 15 + 0.25 + 0.0390625$

$$= 46831.2890625_{10}$$

(e) *Binary coded decimal (BCD).* The binary number system is the simplest and best system for logic circuits and digital computers. However, the denary number system is the most familiar world-wide. Hence, for digital logic circuits and computers to be able to work in the binary system we must have a simple method of converting binary to denary and vice versa. The conventional method using powers of 2 is awkward, and although computers can be instructed to perform the conversions, human operators find it time-consuming to convert long strings of binary digits into a denary number.

The octal and hexadecimal systems are shorthand methods of writing binaries—but they are not a great deal of help in converting them to denary.

To overcome these problems, several binary codes have been devised to *translate each* denary digit into an *equivalent 4-bit binary code* and vice versa, some of which are shown in Fig. 1.1.

After a count of 9_{10}, the BCD *weights* change by a factor equivalent to denary 10.

Denary	Excess-3 (XS-3)	8421 BCD	2421 BCD	7421 BCD
0	0011	0000	0000	0000
1	0100	0001	0001	0001
2	0101	0010	0010	0010
3	0110	0011	0011	0011
4	0111	0100	0100	0100
5	1000	0101	0101	0101
6	1001	0110	0110	0110
7	1010	0111	0111	1000
8	1011	1000	1110	1001
9	1100	1001	1111	1010

Fig. 1.1 Binary coded decimal (BCD) systems.

EXAMPLE 1.5

The denary number 79_{10} may be represented in 8421 BCD as follows:

		0111	1001					
that is,	80	40	20	10	8	4	2	1
	0	1	1	1	1	0	0	1

1.2 Conversions between numbering systems

Digital electronic circuits, microprocessor-based systems, and computers work in binary numbers and binary digits (*bits*). Information is often fed into the above systems in several different forms—including binary, octal, hexadecimal, BCD, alphanumeric, etc.—and it is therefore necessary to be able to convert from one numbering system to another.

Many rules have been established for these conversions, but the following simple procedures can easily be mastered with a little practice:

(a) *Denary-to-binary conversion.* The binary equivalent of a denary number may be found simply by subtracting the value of the highest power of 2—the *most significant digit*, or *bit* (MSB)—from the denary number and writing a 1 in that position. The procedure is repeated on the remainder, progressively, until all binary digits have been formed.

EXAMPLE 1.6

Convert 157_{10} to binary.

$$
\begin{array}{rl}
157\; - & \\
\underline{128} & = \text{highest power of 2} \;\therefore \text{MSB} \;= 1 \\
29\; - & \quad\text{no 64, next significant bit} \;= 0 \\
 & \quad\text{no 32, next significant bit} \;= 0 \\
\underline{16}\; - & \quad 16,\text{ next significant bit} \;= 1 \\
13\; - & \\
\underline{8} & \quad 8,\text{ next significant bit} \;= 1 \\
5\; - & \\
\underline{4} & \quad 4,\text{ next significant bit} \;= 1 \\
\underline{1} & \quad 2,\text{ next significant bit} \;= 0 \\
\underline{1} & \quad 1,\text{ least significant bit} \;= 1 \\
0 &
\end{array}
$$

The binary equivalent of 157_{10} is therefore 10011101_2.

When large denary numbers are to be converted, the above method can lead to errors, and an alternative is continuously to divide by 2 while noting the remainder.

EXAMPLE 1.7

Convert 157_{10} to binary.

```
2 )157
2 )  78    remainder 1  (LSB)
2 )  39    remainder 0
2 )  19    remainder 1
2 )   9    remainder 1
2 )   4    remainder 1
2 )   2    remainder 0
2 )   1    remainder 0
      0    remainder 1  (MSB)
```

The binary equivalent of 157_{10} is therefore 10011101_2.

The above examples show how whole numbers (*integers*) may be converted. When the number to be converted is a *compound* number, i.e., it has a fractional part as well as an integer part, the conversion must be performed in two stages. The integer part is converted separately, as above. Then the fractional part is converted by continuously multiplying the fractional part by 2, the radix of the binary system; the integer part of each multiplication gives the binary digits of the solution.

EXAMPLE 1.8

Converting 25.875 to binary.

Dealing with integer part (25_{10}) first:

```
2 )25
2 )12    remainder 1  (LSB)
2 ) 6    remainder 0
2 ) 3    remainder 0
2 ) 1    remainder 1
    0    remainder 1  (MSB)
```

$$\therefore 25_{10} = 11001_2$$

Secondly, deal with the fractional part (0.875_{10})

0.875 ×	0.75 ×	0.5 ×
2	2	2
1.750	1.50	1.0
↓	↓	↓
1 (MSB)	1	1 (LSB)

$$\therefore 0.875_{10} = 0.111_2$$

Thus, $25.875_{10} = 11001.111_2$.

EXAMPLE 1.9

Convert 158.85_{10} to binary.

Considering the integer part (158_{10}) first:

$$
\begin{array}{r l}
2\,)158 & \\
2\,)\;79 & \text{remainder } 0 = \text{LSB} \\
2\,)\;39 & \text{remainder } 1 \\
2\,)\;19 & \text{remainder } 1 \\
2\,)\;\;9 & \text{remainder } 1 \\
2\,)\;\;4 & \text{remainder } 1 \\
2\,)\;\;2 & \text{remainder } 0 \\
2\,)\;\;1 & \text{remainder } 0 \\
0 & \text{remainder } 1 = \text{MSB}
\end{array}
$$

$\therefore$ the binary equivalent of $158_{10} = 10011110_2$.

Secondly, dealing with the fractional part:

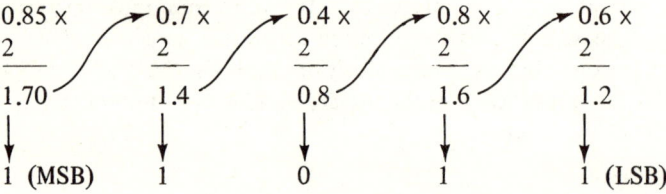

$\therefore$ the binary equivalent of 0.85_{10} is 0.11011_2 to five binary places.

Thus, the binary equivalent of 158.85_{10} is 10011110.11011_2, which is accurate to *five* binary places.

An error is obviously produced in this conversion. This is, of course, always possible and in this case *a true conversion cannot be achieved*. Greater accuracy can be obtained by extending the number of binary places.

(b) *Binary-to-denary conversion*. The denary equivalent of a binary number is most easily determined by adding the equivalent values (or weights) of those digits where a 1 exists:

2^7	2^6	2^5	2^4	2^3	2^2	2^1	2^0	$\cdot$	2^{-1}	2^{-2}	2^{-3}	2^{-4}	2^{-5}
128	64	32	16	8	4	2	1		0.5	0.25	0.125	0.0625	0.03125

EXAMPLE 1.10

Convert 10011101_2 to denary.

$$10011101_2 = 128 + 16 + 8 + 4 + 1$$

$$= 157_{10}$$

Check with Example 1.7.

EXAMPLE 1.11

Convert 11001.111_2 to denary.

$$11001.111_2 = 16 + 8 + 1 + 0.5 + 0.25 + 0.125$$

$$= 25.875_{10} \qquad \qquad \text{Check with Example 1.8.}$$

EXAMPLE 1.12

Convert 10011110.11011_2 to denary

$$10011110.11011_2 = 128 + 16 + 8 + 4 + 2 + 0.5 + 0.25 + 0.0625 + 0.03125$$

$$= 158.84375_{10}$$

Note: If this is checked with Example 1.9 it is possible to determine the error in the denary-to-binary conversion.

(c) *Denary-to-octal conversion*. Similar techniques to denary-to-binary conversion can be applied here. For this conversion, however, we must continuously divide by 8 and note the remainder.

EXAMPLE 1.13

Convert 157_{10} to octal.

$$
\begin{array}{rl}
8\,)\underline{157} & \\
8\,)\underline{\ 19} & \text{remainder 5} = \text{LSD} \\
8\,)\underline{\ \ 2} & \text{remainder 3} \\
0 & \text{remainder 2} = \text{MSD}
\end{array}
$$

$\therefore$ the octal equivalent of 157_{10} is 235_8.

EXAMPLE 1.14

Convert 25.875_{10} to octal.

Converting the whole number (integer) part first:

$$
\begin{array}{rl}
8\,)\underline{25} & \\
8\,)\underline{\ 3} & \text{remainder 1 LSD} \\
0 & \text{remainder 3 MSD}
\end{array}
$$

$\therefore$ the octal equivalent of 25_{10} is 31_8.

Converting the fractional part secondly:

$$
\begin{array}{l}
0.875 \times \\
\underline{8 } \\
7.000 \quad \text{no further remainder} \\
\downarrow \\
\text{MSD}
\end{array}
$$

∴ the octal equivalent of 0.875_{10} is 0.7_8 and the octal equivalent of 25.875_{10} is 31.7_8.

EXAMPLE 1.15

Convert 158.85_{10} to octal.

Converting the integer part first:

$$
\begin{array}{rl}
8\,)158 & \\
8\,)\ \ 19 & \text{remainder } 6 = \text{LSD} \\
8\,)\ \ \ \ 2 & \text{remainder } 3 \\
\ \ \ \ \ 0 & \text{remainder } 2 = \text{MSD}
\end{array}
$$

∴ the octal equivalent of 158_{10} is 236_8.

Considering the fraction part:

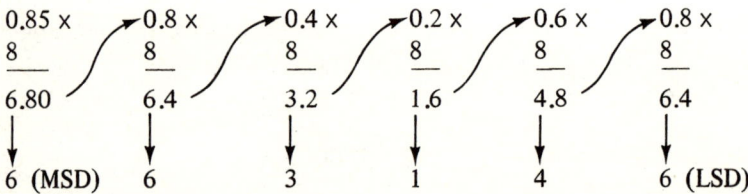

$$
\begin{array}{cccccc}
0.85 \times & 0.8 \times & 0.4 \times & 0.2 \times & 0.6 \times & 0.8 \times \\
8 & 8 & 8 & 8 & 8 & 8 \\
\hline
6.80 & 6.4 & 3.2 & 1.6 & 4.8 & 6.4 \\
\downarrow & \downarrow & \downarrow & \downarrow & \downarrow & \downarrow \\
6\ (\text{MSD}) & 6 & 3 & 1 & 4 & 6\ (\text{LSD})
\end{array}
$$

Therefore, the octal equivalent of 0.85_{10} is 0.663146_8 to six octal places. Note again that a true conversion cannot be achieved. Thus, the octal equivalent of 158.85_{10} is 236.663146_8 to six octal places.

(d) *Octal-to-denary conversion*. The denary equivalent of an octal number is most easily determined by multiplying each digit by its value of power of 8, and then adding:

$$8^3 \quad 8^2 \quad 8^1 \quad 8^0 \cdot \ 8^{-1} \qquad 8^{-2} \qquad\quad 8^{-3} \qquad\quad 8^{-4} \qquad\quad 8^{-5}$$

$$512 \quad 64 \quad 8 \quad 1 \quad\ \ 0.125 \quad 0.015625 \quad 0.0019531 \quad 0.0002441 \quad 0.00003051$$

EXAMPLE 1.16

Convert 31.7_8 to denary.

$$
\begin{aligned}
31.7_8 &= 3 \times 8^1 + 1 \times 8^0 + 7 \times 8^{-1} \\
&= 3 \times 8 + 1 \times 1 + 7 \times 0.125 \\
&= 24 + 1 + 0.875 \\
&= 25.875_{10}
\end{aligned}
$$

(e) *Octal-to-binary conversion*. Since 8 (the octal radix) $= 2^3$, each single octal digit can be represented by a 3-bit binary group. The conversion of an octal number to a binary number is achieved by writing the 3-bit binary equivalent of each octal digit.

EXAMPLE 1.17

Convert 235_8 to binary.

$$\begin{array}{c|c|c} 2 & 3 & 5_8 \\ \hline 010 & 011 & 101_2 \end{array}$$

Thus, the binary equivalent of 235_8 is 10011101_2.

EXAMPLE 1.18

Convert 31.7_8 to binary.

$$\begin{array}{c|c|c} 3 & 1. & 7_8 \\ \hline 011 & 001. & 111_2 \end{array}$$

Thus, the binary equivalent of 31.7_8 is 11001.111_2.

(f) *Binary-to-octal conversion*. The binary number is divided into 3-bit groups, working from the binary point (either to the left or to the right).

EXAMPLE 1.19

Convert 10011101_2 to octal.

$$\begin{array}{c|c|c} 10 & 011 & 101_2 \\ \hline 2 & 3 & 5_8 \end{array}$$

Thus, the octal equivalent of 10011101_2 is 235_8.

EXAMPLE 1.20

Convert 11001.111_2 to octal.

$$\begin{array}{c|c|c} 11 & 001. & 111_2 \\ \hline 3 & 1. & 7_8 \end{array}$$

Thus, the octal equivalent of 11001.111_2 is 31.7_8.

(g) *Denary-to-hex conversion*. This is achieved by continuously dividing by 16 (the radix of the hexadecimal system) and noting the remainder.

EXAMPLE 1.21

Convert 157_{10} to hexadecimal.

$$16 \overline{)157}$$
$$16 \overline{)\ \ 9} \quad \text{remainder D} = \text{LSD}$$
$$0 \quad \text{remainder 9} = \text{MSD}$$

Thus, the hex equivalent of 157_{10} is 9 DH.

EXAMPLE 1.22

Convert 25.875_{10} to hexadecimal.
    Converting the integer part first:

$$16 \overline{)25}$$
$$16 \overline{)\ \ 1} \quad \text{remainder 9} = \text{LSD}$$
$$0 \quad \text{remainder 1} = \text{MSD}$$

The hex equivalent of 25_{10} is 19H.
    Converting the fractional part:

$$0.875 \times$$
$$\underline{16}$$
$$14.000 \quad \text{no further digits}$$
$$\downarrow$$
$$E \quad (\text{MSD})$$

The hex equivalent of 0.875_{10}, is 0.EH and the hex equivalent of 25.875_{10} is 19.EH.

(h) *Hex-to-denary conversion*. The denary equivalent of a hex number is most easily determined by multiplying each digit by its value of power of 16 and then adding:

$$16^1 \quad 16^0 \cdot 16^{-1} \qquad 16^{-2}$$
$$16 \quad 1 \quad 0.0625 \quad 0.00390625$$

EXAMPLE 1.23

Convert 9 DH to denary.

$$9\text{DH} = 9 \times 16^1 + \text{D} \times 16^0$$
$$= 9 \times 16 + 13 \times 1$$
$$= 144 + 13$$
$$= 157_{10}$$

EXAMPLE 1.24

Convert 19.EH to denary.

$$19.\text{EH} = 1 \times 16^1 + 9 \times 16^0 + \text{E} \times 16^{-1}$$
$$= 1 \times 16 + 9 \times 1 + 14 \times 0.0625$$
$$= 25.875_{10}$$

(i) *Hex-to-binary conversion.* Since 16 (the hex radix) = 2^4, each single hex digit can be represented by a 4-bit binary group. The conversion of a hex number to a binary number is achieved by writing the 4-bit binary equivalent of each hex digit.

EXAMPLE 1.25

Convert 9DH to binary.

$$9 \qquad \text{D} \quad \text{H}$$
$$1001 \qquad 1101_2$$

Thus, the binary equivalent of 9DH is 10011101_2.

EXAMPLE 1.26

Convert 19.EH to binary.

$$1 \qquad 9. \qquad \text{E} \quad \text{H}$$
$$0001 \qquad 1001. \qquad 1110_2$$

Thus, the binary equivalent of 19.EH is 11001.111_2.

(j) *Binary-to-hex conversion.* The binary number is divided into 4-bit groups, working from the binary point (either to the left or to the right).

EXAMPLE 1.27

Convert 10011101_2 to hex.

$$1001 \qquad 1101_2$$
$$9 \qquad \text{D} \quad \text{H}$$

Thus, the hex equivalent of 10011101_2 is 9DH.

EXAMPLE 1.28

Convert 11001.111_2 to hex.

$$0001 \qquad 1001 \quad . \quad 1110_2$$
$$1 \qquad 9 \quad . \quad \text{E} \quad \text{H}$$

Thus, the hex equivalent of 11001.111_2 is 19.EH.

1.3 Signed numbers

When dealing with arithmetic problems it is necessary to be able to represent whether the numbers are positive or negative. Although this does not present any difficulty for humans and paper exercises, it *does* present a significant problem for electronic logic circuits and computers.

A *signed number* is a binary number which uses an additional bit—usually the left-hand (or most significant) digit—to indicate the *sign* of the number, and is called the *sign bit*. The convention usually adopted is that the sign bit is 0 for a positive number and 1 for a negative number. The remaining digits represent the magnitude of the number.

EXAMPLE 1.29

Assuming that *eight* bits can be accommodated by a digital system, write down the signed binary numbers for: (a) $+122_{10}$; (b) -67_{10}.

(a) $+122_{10}$ = 01111010_2

　　　　　　　　　sign bit

(b) -67_{10} = 11000011_2

　　　　　　　　　sign bit

The signed number system imposes the severe limitation of reducing the maximum number representation by half, e.g., an 8-bit unsigned binary number can represent numbers up to $+255_{10}$, whereas using a sign bit the maximum number falls to $+127_{10}$. However, using this method, we can also represent numbers down to -127_{10}.

Further difficulties are encountered in the logic circuitry necessary to handle signed numbers.

1.4 Floating point

All the numbers written so far have used *fixed-point* notation, i.e., the 'radix point' is fixed at the point where the power of the radix changes from positive to negative. When we carry out complex calculations such as multiplication, or wish to represent very large or very small numbers, it becomes necessary to keep track of the radix point in order to determine its position in the solution. While this may not be too difficult to deal with on paper, it is more of a problem for electronic digital systems. Digital systems, therefore, often use a *floating-point* notation when dealing with arithmetic processes. The floating-point notation is a technique which minimizes the logic circuitry required to determine the position of the radix point, and represents a number in two parts, the *mantissa* and the *exponent*.

EXAMPLE 1.30

Write down a selection of floating-point notations for the following fixed-point numbers: (a) 158.85_{10}; (b) 31.7_8; (c) 11001.111_2; (d) 19.EH.

(a) 158.85_{10} = 158.85×10^0 = 15.885×10^1 = 1.5885×10^2 = 0.5885×10^3
(b) 31.7_8 = 31.7×8^0 = 3.17×8^1 = 0.317×8^2
(c) 11001.111_2 = 11001.111×2^0 = 1100.1111×2^1 = 110.01111×2^2
$$= 11.001111 \times 2^3 = 1.1001111 \times 2^4 = 0.11001111 \times 2^5$$
(d) 19.EH = $19.E \times 16^0$ = $1.9E \times 16^1$ = $0.19E \times 16^2$

Floating-point binary numbers can also use the sign bit form of representation. It should be noted that *both* the mantissa *and* the exponent must have sign bits, since they are required to represent both positive and negative numbers. In general, the mantissa is a fraction and the binary point is placed between the sign bit and the next mantissa digit. The exponent is a positive or negative integer.

1.5 Modulus

The *modulus* of a binary number system is the denary equivalent which that number of binary digits can represent. Thus, a 4-bit binary counter is called a *modulo-16* binary counter, and a 3-bit binary counter is called a *modulo-8* binary counter. Therefore, in the case of a straight binary counter or system, the modulus is given by the radix (2) raised to the power of the number of bits. For example, for the 4-bit counter, modulus = 2^4 = 16_{10}.

There are many requirements for counters having modulus numbers other than those produced from 'powers of 2'. These can all be achieved from basic binary counters.

1.6 Binary arithmetic

The four most common arithmetic operations (addition, subtraction, multiplication, and division) can all be performed by manipulations of *addition*. It is for this reason that addition is *the* most important arithmetic operation.

The binary numbering system, as mentioned above, forms the basis of 'machine code', i.e., the form in which digital electronics and computers work. Irrespective of how data is input to the machine or system, it must be converted into a binary code for the machine to understand it and respond to it. In all numbering systems it is important to line up the radix point before proceeding.

(a) *Binary addition.* The same basic mathematical processes are involved in the addition of two numbers regardless of the radix of the numbering system.
   Starting with the least significant bit (LSB), add the 1s in the usual way:

   (i) $0 + 0 = 0$;
   (ii) $0 + 1$ and $1 + 0 = 1$;
   (iii) $1 + 1 = 0$, carry 1 to the next most significant column.

EXAMPLE 1.31

Augend	11011101_2 +	$\equiv$	221_{10} +
Addend	100111_2 +	$\equiv$	39_{10}
Carry	11111111		1
Sum	100000100_2		260_{10}

EXAMPLE 1.32

Augend	1010.011_2 +	$\equiv$	10.375_{10} +
Addend	111.11_2	$\equiv$	7.75_{10}
Carry	$11111\ 1$		$1\ 1$
Sum	10010.001_2	$\equiv$	18.125_{10}

(b) *Binary subtraction*. In subtraction the *subtrahend* is subtracted from the *minuend* to give the *difference*. Again the method is the same as that used in other numbering systems, except that in this case the rules are simpler:

  (i) $0 - 0$, and $1 - 1 = 0$;
  (ii) $1 - 0 = 1$;
  (iii) $0 - 1 = 1$, borrow 1 from next most significant column.

EXAMPLE 1.33

Minuend	101000_2 −	$\equiv$	40_{10} −
Subtrahend	10110_2	$\equiv$	22_{10}
Borrow	$1\ \ 11$		
Difference	10010_2	$\equiv$	18_{10}

EXAMPLE 1.34

Minuend	10010_2 −	$\equiv$	18_{10}
Subtrahend	11100_2	$\equiv$	28_{10}
Borrow	1 11		
Difference	10110_2 −	$\equiv$	-10_{10}
	1		
	1		
Final result	10101_2		

In this case, the subtrahend is larger than the minuend, so yielding a negative difference (-10_{10}). Using the rules of subtraction given above the binary result 10110_2 is obtained and an *additional borrow digit* appears.

If now this additional borrow digit is subtracted from the first difference (10110_2), as shown above, then the *final result 10101_2 is related* to the expected result (-10_{10}). This final result is in the *1s complement form*. If all the 1s are changed to 0s, and all the 0s changed to 1s, i.e., if all digits in the 1s complement number are inverted, then the binary number 1010_2 is produced, which is equivalent to 10_{10}.

The above method of dealing with subtraction may be satisfactory as a paper exercise, but to provide the logic circuitry to deal with it in a machine (e.g., a computer) presents some difficulties. If it is assumed that the machine can only add, then the arithmetic operation of subtraction must be achieved by *complementary addition*. This technique can, of course, be applied in any numbering system. Since digital electronic systems and computers work with *binary* numbers, subtraction is achieved by forming the *2s complement* of the subtrahend and then adding this to the minuend.

Binary subtraction must be carried out in four stages:

1. Ensure that both numbers have the same number of digits and/or the same as the digital circuitry.
2. Form the *1s complement* of the subtrahend by changing all 1s to 0s and all 0s to 1s.
3. Form the *2s complement* of the subtrahend by adding 1 to the LSB of the 1s complement.
4. Add the 2s complement of the subtrahend to the minuend.

EXAMPLE 1.35

Using the binary numbering system, subtract 22_{10} from 47_{10} by the method of complementary addition.

Assume that the equipment to be used can accommodate eight binary digits (i.e., eight bits).

$$
\begin{array}{llllll}
\text{Minuend} & = & 47_{10} & \equiv & 101111_2 & = & 00101111_2 \\
\text{Subtrahend} & = & 22_{10} & \equiv & 10110_2 & = & 00010110_2
\end{array}
$$

$$
\begin{array}{lll}
\text{1s complement of subtrahend} & = & 11101001 \\
\text{Plus 1} & & \underline{\hspace{1em}+1}
\end{array}
$$

$$
\therefore \text{2s complement of subtrahend} = 11101010
$$

$$
\begin{array}{lll}
\text{Finally} \quad \text{Minuend} & & 00101111_2 \ + \\
\quad\quad\quad\text{2s complement of subtrahend} & & \underline{11101010_2} \\
\\
\quad\quad\quad\text{Carry} & & \underline{111\ \ 111} \\
\\
\quad\quad\quad\text{Sum} & & 100011001_2
\end{array}
$$

Note that the carry out of the MSB of the results will be lost, therefore the solution is $00011001_2 = 25_{10}$. The presence of a carry out digit indicates that the result is *positive*.

EXAMPLE 1.36

$$1010.011 -$$
$$111.11$$

Subtrahend = 0111.110 to have the same number of digits as the minuend.

1s complement of subtrahend =	1000.001
Plus 1	+1
∴ 2s complement of subtrahend =	1000.010
Finally, add	1010.011 +
	1000.010
	10010.101

The carry out of the MSB indicates that the result is positive, therefore the solution is 0010.101_2 or 10.101_2.

EXAMPLE 1.37

$$10010_2 -$$
$$11100_2$$

Subtrahend = 11100_2.

1s complement of subtrahend =	00011
Plus 1	+1
∴ 2s complement of subtrahend =	00100
Finally add	10010
	00100
	10110

Note that this result does not produce a carry out, and the result is therefore *negative*. This will always be the case when the subtrahend is greater than the minuend. The result is in 2s complement form, so that the final solution may be determined:

$$10110 = -1 \times 2^4 + 1 \times 2^2 + 1 \times 2^1$$
$$= -16 + 4 + 2$$
$$= -16 + 6$$
$$= -10_{10}$$

or

$$\text{2s complement of solution} \quad = \quad 10110$$

$$-1$$

Borrow   1

↑s complement of solution   10101

∴ solution $= \ -01010_2 \ \equiv \ -10_{10}$

(c) *Binary multiplication*. The operands in multiplication are the *multiplicand* and *multiplier* which are multiplied together to produce the *product*.

The simplest method of dealing with multiplication is to add the multiplicand to itself as many times as defined by the multiplier. However, when dealing with binary numbers, we only have to use 1s and 0s, so that the process known as 'long multiplication' becomes a process referred to as *shift-and-add*, and is the method used by most computer systems.

Rules for binary multiplication are simply:

(i) 1×1 $= 1$;
(ii) 1×0 and $0 \times 1 = 0$;
(iii) 0×0 $= 0$.

Due to these simple rules, each non-zero partial product is simply the multiplicand shifted by the number of bits of significance of the multiplier bit which produced it.

EXAMPLE 1.38

Perform the multiplication of 55_{10} and 11_{10} in binary.

Multiplicand	110111_2 ×	≡	55_{10} ×	
Multiplier	1011_2	≡	11_{10}	
Partial products	110111		55	
	110111		55	
	000000			
	110111			
Carry	1111111			
	1		1	
Product	1001011101_2	≡	605_{10}	

EXAMPLE 1.39

Multiplicand	1011.010_2 ×	≡	11.250_{10} ×
Multiplier	11.101_2	≡	3.625_{10}

	1011 010	56 250
Partial products	101101 0	225
	1011010	6750
	1011010	33750

Carry	11111111	1111
	111	

Product	$101000.110\ 010_2$	≡	$40.781\ 250_{10}$

Note that it is common practice to omit the partial products produced by zero bits of the multiplier.

(d) *Binary division*. The operation is that the *dividend* is divided by the *divisor* to produce the *quotient* and *remainder*. Division of binary numbers is best achieved by the *shift-and-subtract* method which is, in fact, similar to the familiar technique of 'long division'.

EXAMPLE 1.40

Divide 605_{10} by 11_{10} in binary.

$$\text{Dividend} = 1001011101_2 \equiv 605_{10}$$
$$\text{Divisor} \quad = \quad\quad\quad 1011_2 \equiv \quad 11_{10}$$

```
              110111₂   (quotient)                    55₁₀    (quotient)
     1011₂ )1001011101₂  (dividend)           11₁₀ )605₁₀
      −1011                                          55
        1111                                         55
      −1011                                          55
        10011                              Remainder  00
       −1011
         10000
        −1011
          1011
        −1011
Remainder   0000
```

EXAMPLE 1.41

Divide 40.78125_{10} by 3.625_{10} in binary.

$$
\begin{aligned}
\text{Dividend} &= 101000.11001_2 &= 40.78125_{10} \\
\text{Divisor} &= 11.101_2 &\equiv 3.625_{10}
\end{aligned}
$$

Now, as in denary division, it is common practice to make the divisor an integer. This can be achieved by moving the binary point to the right. However, it must be noted that the denary equivalents of the binary numbers thus formed will be different from the originals.

$$
\begin{aligned}
\text{Divisor} &= 11101_2 &\equiv 29_{10} \\
\text{Dividend} &= 101000110.01_2 &\equiv 326.25_{10}
\end{aligned}
$$

```
        1011.01₂  (quotient)                    11.25₁₀
11101 )101000110.01₂                      29 )326.25₁₀
       11101                                   29

       101111                                  36
       11101                                   29

       100100                                  72
       11101                                   58

        11101                                 145
        11101                                 145

Remainder   00000                    Remainder    0
```

Another means of dealing with binary division is the *restoring method*. This is similar to the long division used above, except that the divisor and dividend digits are compared by subtraction, and the quotient is formed depending on whether or not the subtraction was successful.

If it is assumed that the divisor is a 4-bit binary number, and that it is larger than the most significant four digits of the dividend, then the first subtraction is unsuccessful, i.e., it produces a negative result, which can be indicated by the presence of a 'borrow' digit. In this case, the first digit in the quotient will be 0, and it is now necessary to *restore* the dividend to its original value, which is achieved by *adding* the divisor. If the result of the first subtraction is successful, i.e., the result is positive, then the first digit in the quotient will be 1.

The next digit of the dividend is 'brought down' as before, and the divisor is shifted one place to the right ready for the next subtraction. This process is repeated until all digits of the dividend have been used.

EXAMPLE 1.42

Use the restoring method of division to solve the following problem:

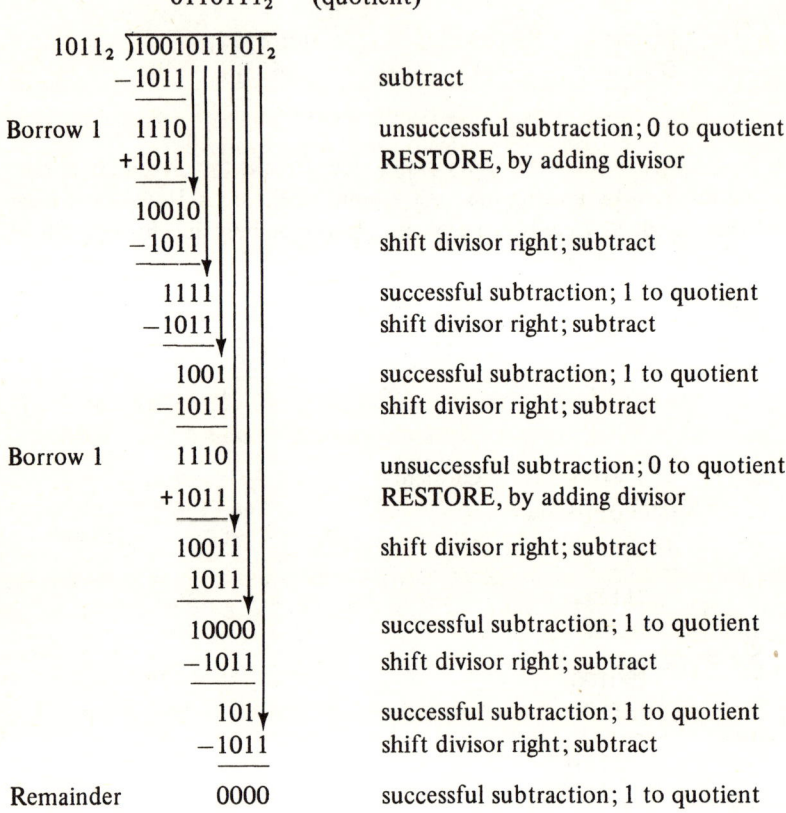

$$0110111_2 \quad \text{(quotient)}$$

1011_2 $)\overline{1001011110 1_2}$	
-1011	subtract
Borrow 1 1110	unsuccessful subtraction; 0 to quotient
$+1011$	RESTORE, by adding divisor
10010	
-1011	shift divisor right; subtract
1111	successful subtraction; 1 to quotient
-1011	shift divisor right; subtract
1001	successful subtraction; 1 to quotient
-1011	shift divisor right; subtract
Borrow 1 1110	unsuccessful subtraction; 0 to quotient
$+1011$	RESTORE, by adding divisor
10011	shift divisor right; subtract
1011	
10000	successful subtraction; 1 to quotient
-1011	shift divisor right; subtract
101	successful subtraction; 1 to quotient
-1011	shift divisor right; subtract
Remainder 0000	successful subtraction; 1 to quotient

The solution is thus 110111_2 which can be compared with the solution to Example 1.40.

Yet another variation of division is the *non-restoring method*, which allows faster execution when carried out by computers. The non-restoring method of division is performed by subtracting the divisor from the dividend using the 2s complementary addition (after aligning the MSBs) and noting the value of the sign bit of the difference produced:

1. If the sign bit = 1 (positive result), write a 0 in the partial quotient. Shift the difference and the partial quotient one position to the left and add the divisor to the difference.
2. If the sign bit = 0 (negative result), write a 1 in the partial quotient. Shift the difference and the partial quotient one position to the left and subtract the divisor from the difference.

This process is repeated to produce any order of accuracy.

EXAMPLE 1.43

Using the non-restoring method, divide 110111_2 by 1011_2.

Assuming that both divisor and dividend are positive numbers:

	Divisor				Partial quotient
(0)	1011	(0) 110111	(dividend)		
sign bit		(1) 0101		2s complement of divisor	
Carry 1	(0) 001011			sign bit = 0, write 1 in LSB	1
	(0) 010110			shift left	1
	(1) 0101			subtract divisor	
	(1) 10101			sign bit = 1, write 0 in LSB	10
	(1) 0101			shift left	10
	(0) 1011			add divisor	
Carry 1	(0) 0000			sign bit = 0, write 1 in LSB	101
	(0) 000			shift left	101
	(1) 0101			subtract divisor	
	(1) 0101			sign bit = 1, write 0 in LSB	1010
	(1) 101			shift left	1010
	(0) 1011			add divisor	
Carry 1	(1) 0101			sign bit = 1, write 0 in LSB	10100

The position of the binary point may be determined by examination of the dividend and divisor. In this case, the dividend contains *six* digits, therefore its MSB is 2^5, while the divisor contains *four* digits and its MSB is 2^3. Therefore the result of the division is $2^{5-3} = 2^2$, i.e., *three* digits. Thus, by examination of the partial quotient, it can be seen that the true quotient = 101.00_2 or 101_2 in this case.

EXERCISES TO CHAPTER 1

1. Convert the following denary numbers into binary:

   (a) 17_{10}, (b) 35_{10}, (c) 37.0625_{10}, (d) 28.625_{10}, (e) 221.75_{10}.

2. Convert the following denary numbers into octal:

   (a) 11_{10}, (b) 23_{10}, (c) 28.875_{10}, (d) 133.046875_{10},
   (e) 251.1875_{10}.

3. Convert the following denary numbers into hexadecimal:

   (a) 9_{10}, (b) 14_{10}, (c) 73.875_{10}, (d) 255.625_{10} (e) 132.125_{10}.

4. Convert the following binary numbers into denary:

   (a) 1011_2, (b) 11011_2, (c) 110.011_2, (d) 10111.1001_2,
   (e) 11010111.00111_2.

5. Convert the following octal numbers into denary:

   (a) 21_8,   (b) 37_8,   (c) 3.75_8,   (d) 133.35_8,   (e) 275.12_8.

6. Convert the following hexadecimal numbers into denary:

   (a) 9H,   (b) EBH,   (c) 4F.3AH,   (d) F3.CBH,   (e) A4.B3H.

7. Convert the following octal numbers to binary:

   (a) 35_8,   (b) 763_8,   (c) 65.732_8,   (d) 357.214_8,   (e) 477.653_8.

8. Convert the following hexadecimal numbers to binary:

   (a) 17H,   (b) C3H,   (c) 3E.7AH,   (d) 76.AFH,   (e) F.E4H.

9. Convert the following binary numbers to octal:

   (a) 101_2,   (b) 10011_2,   (c) 10111.101_2,   (d) 1010.01011_2,
   (e) 11011101.1011_2.

10. Convert the following binary numbers to hexadecimal:

    (a) 1001_2,   (b) 10101_2,   (c) 110110.1101_2,   (d) 1010111.11011_2,
    (e) 11110111.111_2.

11. Add the following pairs of numbers in binary.

    (a) $11000101_2 + 101101_2$ (b) $63_{10} + 27_{10}$
    (c) $23_8 + 33_8$ (d) CEH + 6AH
    (e) $111.001_2 + 1000.1_2$ (f) $37.0625_{10} + 28.625_{10}$
    (g) $3.77_8 + 21.15_8$ (h) 47.5H + A0.25H

12. Solve the following subtractions in binary, using *both* the arithmetic method
    and complementary addition:

    (a) $11101110_2 - 11011100_2$,   (b) $28_{10} - 25_{10}$,   (c) $35_8 - 13_8$,
    (d) B8H − 6DH,   (e) $1011011.01101_2 - 100100.11_2$,
    (f) $16.375_{10} - 8.75_{10}$,   (g) $27.32_8 - 15.17_8$,   (h) FE.FH − A6.2H.

13. Multiply the following pairs of numbers in binary:

    (a) $7_{10} \times 12_{10}$ (b) $13_8 \times 11_8$
    (c) $1001111_2 \times 1100_2$ (d) F3H × A2H
    (e) $123.5_{10} \times 15.25_{10}$ (f) $137.5_8 \times 17.25_8$
    (g) $11110101.101_2 \times 11011.11_2$ (h) EF.3BH × C3.AH

14. Divide the following pairs of numbers in binary:

    (a) $72_{10} \div 8_{10}$ (b) $63_8 \div 21_8$
    (c) $110110_2 \div 1001_2$ (d) C8H ÷ 19H
    (e) $106.09375_{10} \div 2.125_{10}$ (f) $26.2_8 \div 3.4_8$
    (g) $11011.101_2 \div 110.1_2$ (h) EB.C3H ÷ F.AH

2 Boolean algebra and logic elements

2.1 Introduction

In 1854 George Boole wrote a paper entitled 'An Investigation into the Laws of Thought'. Normal mathematics, although extremely useful in many intellectual pursuits, cannot deal with every aspect of *thought*. The following example shows the inadequate nature of normal algebra:

Suppose the following two statements are made—

Cats are animals

Dogs are animals

The algebraic conclusion could be 'therefore cats are dogs', a conclusion which is obviously absurd. However, suppose we represent the above statements by normal algebraic expressions:

$$A = B$$
$$C = B$$
$$\therefore A = C$$

which is a perfectly valid deduction.

The inherent difficulty is a question of *language*. The mathematical sign of equality is used to represent the word 'are'. The fact that a cat is an animal *is* true, but it is *not* true to say that it 'equals' an animal, since it is only a *sub*-class of the much larger general class of animals.

Boole developed an entirely new system and called it 'The Algebra of Classes'. Apart from its interest to the mathematical geniuses of the age, Boole's treatise rotted away in the corners of the world's libraries until 1938. By that time, telephone and communication engineering had reached a high degree of complexity and Boole's methods suddenly became famous, due to a paper entitled 'Symbolic Analysis of Relay and Switching Circuits' published by C. E. Shannon.

Shannon discovered that Boole's 'Algebra of Classes' was a powerful tool with which to analyse and represent complicated circuitry employing 'two-state' ideas.

2.2 Basic rules of boolean algebra

1. A quantity can have only one of two possible values; it can be a '1' or a '0'. No other value exists.

2. The usual meanings of certain mathematical signs take on entirely different meanings as follows:

$A \cdot B$ means 'A AND B', not A times B (logical product)
$A + B$ means 'A OR B', not A plus B (logical sum)
$\overline{A}$ means 'NOT A' or 'the complement of A'

3. The sign of equality (=) has a new significance, and may best be defined as follows:

= means   'an output exists' or
'the switch is closed' or
'the function is valid'

2.3 Basic logic gates and truth tables

Devices used in logic systems to control the flow of *information* are known as *logic gates* which are opened and closed by the application of logic signals at their inputs. The basic range of logic elements are AND, OR, and NOT, as indicated above.

Modern electronics, however, has led to the development and wide availability of two further logic gates, the NAND and NOR gates, which are combinations of the basic logic functions.

The graphical symbols used in pure logic diagrams represent thought processes, and are therefore independent of the equipment which might be used to implement them, i.e., the same symbols are common to all methods of implementation, whether electronic, pneumatic, hydraulic, or mechanical, although in this text we shall be dealing with electronic equipment only.

Many different systems of symbolization have been used in equipment drawings and publications throughout the world. Early British Standards Specifications used circles for all the basic gates, with inhibition shown by a short perpendicular line drawn across the data flow line. The revision in 1969 of BS 3939, Section 21—Logic Symbols, led to the D outline symbols being widely used in the United Kingdom and particularly in the sphere of education and training. This Standard was revised in July 1977, and recommends the use of rectangular outlines. However, due to the lead that the United States of America have established in the component and equipment manufacture and application, the US Military Standard has been most widely used throughout the world. This has become 'the industry-standard' system, and in consequence symbols of that standard only are used throughout this text. Alternative standard symbols are shown in Appendix A.

(a) *The AND gate.* The logic function AND can be represented by the simple series electrical switching circuit shown in Fig. 2.1(*a*), in which the lamp will only become illuminated when both switches *A and B* are closed. This function may be represented by the logic symbol shown in Fig. 2.1(*b*), and for which the boolean equation is:

$$F = A \cdot B \tag{2.1}$$

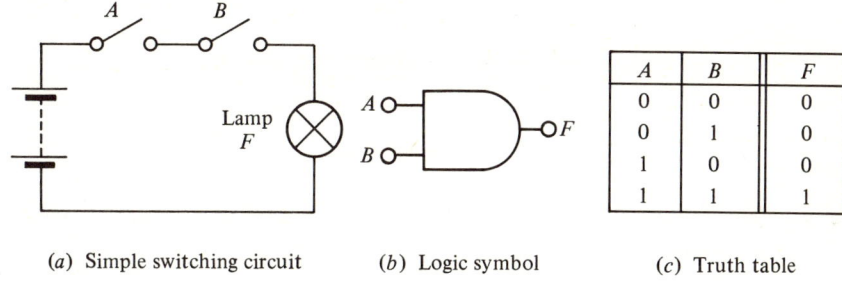

(a) Simple switching circuit   (b) Logic symbol   (c) Truth table

Fig. 2.1. The AND gate.

This system has only *two* inputs, which means that there are *four* possible combinations of signals which can be applied to the inputs. In general, the number of possible combinations of inputs to a system is given by 2^N, where N is the number of inputs to the system.

The *truth table* is a table which shows the state of the output for all possible combinations of inputs, as shown for the two-input AND gate in Fig. 2.1(c).

(b) *The OR gate.* The logical function OR can be represented by the simple parallel electrical switching circuit shown in Fig. 2.2(a), in which the lamp

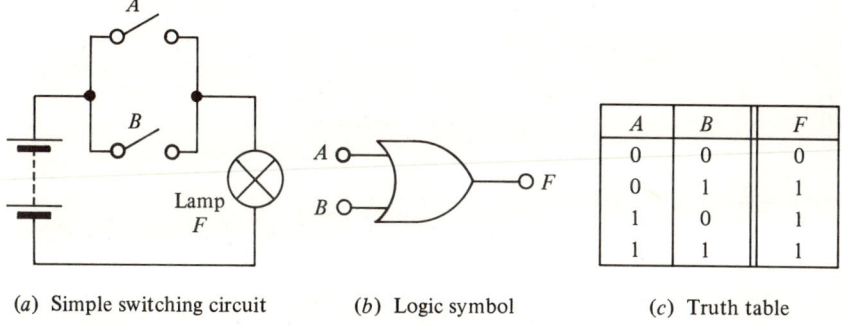

(a) Simple switching circuit   (b) Logic symbol   (c) Truth table

Fig. 2.2. The OR gate.

will be illuminated upon the closure of either switch *A or B or* both switches *A and B*. The logical OR function may be represented by the logic symbol shown in Fig. 2.2(b), and for which the boolean equation is:

$$F = A + B \tag{2.2}$$

The truth table for the logical OR gate is shown in Fig. 2.2(c).

(c) *The NOT gate.* The logic symbol for the NOT logic function is shown in Fig. 2.3. Note the small circle indicating the inhibiting action. Alternatively, this small circle can be shown at the input of the symbol and not at the output.

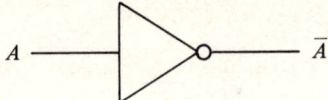

Fig. 2.3. The NOT gate.

The output signal of the NOT gate will be 1 when the input signal is 0, and vice versa, i.e., the input signal is *complemented*.

(d) *The NAND gate (NOT-AND).* If the output of the AND gate is fed to the input of the NOT gate, then the NOT-AND, or NAND logic function is produced. The logic symbol for the NAND gate is shown in Fig. 2.4(*a*), and the boolean equation is:

$$F = \overline{A \cdot B} \tag{2.3}$$

The truth table for this two-input NAND gate is shown in Fig. 2.4(*b*).

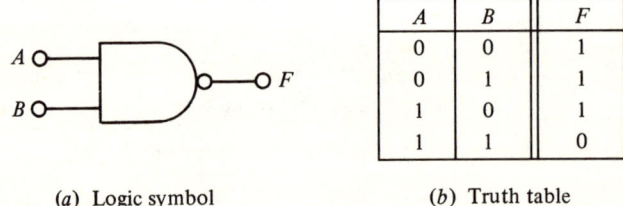

A	*B*	*F*
0	0	1
0	1	1
1	0	1
1	1	0

(*a*) Logic symbol   (*b*) Truth table

Fig. 2.4. The NAND gate.

(e) *The NOR gate (NOT-OR).* If the output of the OR gate is fed to the input of the NOT gate, then the NOT-OR, or NOR logic function is produced. The logic symbol for the NOR gate is shown in Fig. 2.5(*a*), and the boolean equation is:

$$F = \overline{A + B} \tag{2.4}$$

The truth table for this two-point NOR gate is shown in Fig. 2.5(*b*).

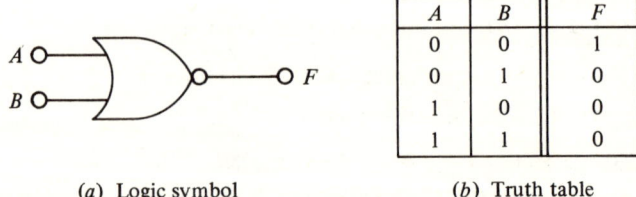

A	*B*	*F*
0	0	1
0	1	0
1	0	0
1	1	0

(*a*) Logic symbol   (*b*) Truth table

Fig. 2.5. The NOR gate.

2.4 The laws of logic

The truth of many logic statements is self-evident, while that of others may not be so clear. Provided the statement is accurate, it is possible to test its truth.

Using binary notation, we say that if a statement is true, i.e., if the function exists, it has a value of 1. If it is false, it does not exist, and has a value of 0.

The basic laws of logic are frequently referred to as *identities*, and are listed as follows:

(a) $A.0 = 0$
(b) $A + 0 = A$
(c) $A.1 = A$
(d) $A + 1 = 1$
(e) $A.A = A$
(f) $A + A = A$
(g) $A.\bar{A} = 0$
(h) $A + \bar{A} = 1$

The identities listed above may be verified by using the simple switching arrangements shown in Fig. 2.6, in which AND functions are represented by series-connected switches and OR functions are represented by parallel-connected switches.

In addition to the boolean identities listed above, if a complemented signal is applied to a NOT gate, the output of that gate will be the original signal:

(i) $\bar{\bar{A}} = A$

Several boolean theorems have been established as 'laws of logic' to enable the simplification of boolean expressions and logic networks:

(j) $A.B = B.A$ ⎫
(k) $A + B = B + A$ ⎬ commutative

(l) $A.(B.C) = (A.B).C$ ⎫
(m) $A + (B + C) = (A + B) + C$ ⎬ associative

(n) $A.(B + C) = A.B + A.C$ ⎫
(o) $A + (B.C) = (A + B).(A + C)$ ⎬ distributive

(p) $\overline{A.B.C} = \bar{A} + \bar{B} + \bar{C}$ ⎫
(q) $\overline{A + B + C} = \bar{A}.\bar{B}.\bar{C}$ ⎬ De Morgan's dual

The distributive expressions may be compared to the normal mathematical processes of *factorization* and *expansion*.

When writing down the evaluating boolean equations, generous use of brackets helps to clarify the meanings of the expressions.

2.5 Boolean manipulations, truth tables, and logic networks

The applications of the techniques of boolean algebra, truth tables, and logic networks are illustrated in the following examples:

EXAMPLE 2.1

Simplify $F = A + B + 1$
            $F = (A + B) + 1$

Substitute $(A + B)$ for A in identity 2.4(d)

$$\therefore F = 1$$

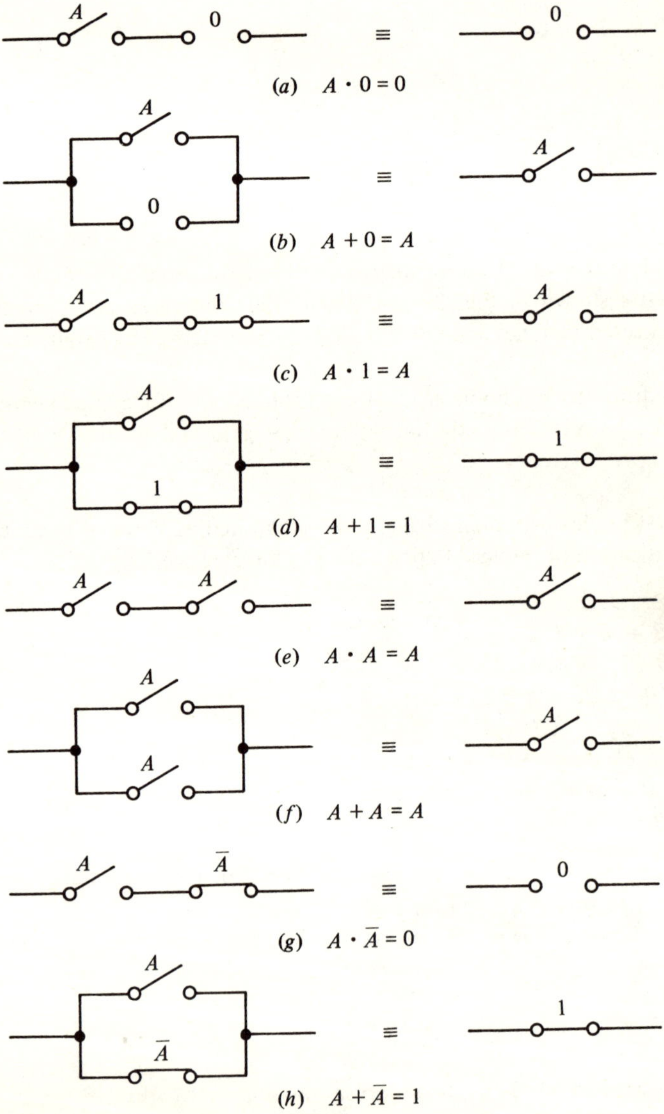

(a) $A \cdot 0 = 0$

(b) $A + 0 = A$

(c) $A \cdot 1 = A$

(d) $A + 1 = 1$

(e) $A \cdot A = A$

(f) $A + A = A$

(g) $A \cdot \bar{A} = 0$

(h) $A + \bar{A} = 1$

Fig. 2.6. Basic laws of logic.

EXAMPLE 2.2

Simplify
$$F = A + B + 0$$
$$F = (A + B) + 0$$

Substitute $(A + B)$ for A in identity 2.4(b)

$$\therefore F = A + B$$

EXAMPLE 2.3

Simplify
$$F = (A + \bar{B}.C) . (A + \bar{B}.C)$$

Substitute $(A + \bar{B}.C)$ for A in identity 2.4(e)

$$\therefore F = (A + \bar{B}.C)$$

EXAMPLE 2.4

Simplify
$$F = A.B + \overline{A.B}$$

Substitute $(A.B)$ for A in identity 2.4(h)

$$\therefore F = 1$$

EXAMPLE 2.5

Simplify
$$F = A + B.0$$
$$F = A + (B.0)$$

Substitute $(B.0)$ for 0 as in identity 2.4(a)

$$\therefore F = A$$

EXAMPLE 2.6

Use a truth table to prove that the boolean expression $A + A.B = A$ is true.

A truth table gives the output conditions of a logic network or system for all combinations of its inputs. The input variables in this case are A and B, and first we write down all the four possible combinations of A and B. We can then derive the logic states for $A.B$ and progressively build up to $A + A.B$ (the original expression).

The truth table is shown in Fig. 2.7.

A	B	$A \cdot B$	$A + A \cdot B$
0	0	0	0
0	1	0	0
1	0	0	1
1	1	1	1

Fig. 2.7. Truth table for Example 2.6.

Examination of the truth table reveals that the final column for $A + A.B$ is identical to the column for A. Therefore $A + A.B = A$ is true.

EXAMPLE 2.7

Use the truth table to prove that $A.(B + C) = A.B + A.C$ is true.

Applying the same techniques as in Example 2.6, we construct the truth table, building up the terms. In this case, there are three input variables, as shown in the truth table in Fig. 2.8.

Examination of the truth table reveals that the final two columns are identical. Therefore, $A.(B + C) = A.B + A.C$ is true.

A	B	C	$(B + C)$	$A \cdot B$	$A \cdot C$	$A \cdot (B + C)$	$A \cdot B + A \cdot C$
0	0	0	0	0	0	0	0
0	0	1	1	0	0	0	0
0	1	0	1	0	0	0	0
0	1	1	1	0	0	0	0
1	0	0	0	0	0	0	0
1	0	1	1	0	1	1	1
1	1	0	1	1	0	1	1
1	1	1	1	1	1	1	1

Fig. 2.8. Truth table for Example 2.7.

EXAMPLE 2.8

Draw the logic network (using basic gates) for the boolean equation

$$F = (A + B).C$$

Now, in this case, we can represent the boolean equation in either of two ways:

$$F = (A + B).C$$
or
$$F = A.C + B.C$$

These two equations can be easily represented by logic networks as shown in Fig. 2.9.

Examination of the logic networks shown in Fig. 2.9 reveals that the network shown in Fig. 2.9(a) for $F = (A + B).C$ is the simplest form, since only two logic gates are required.

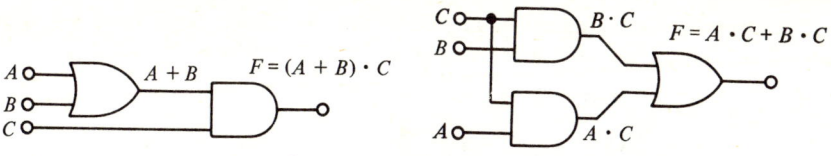

(a) Logic network for $F = (A + B) \cdot C$ (b) Logic network for $F = B \cdot C + A \cdot C$

Fig. 2.9. Logic networks for Example 2.8.

EXAMPLE 2.9

(a) Draw the logic network (using basic logic gates) for the boolean equation: $F = \overline{A.B + \overline{B.C}}$.

(b) Construct the truth table for the above boolean equation. Hence simplify the function and draw the logic network for the simplified logic function.

(a) The logic network for $F = \overline{A.B + \overline{B.C}}$ is drawn in Fig. 2.10.

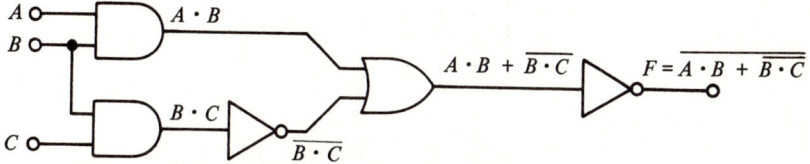

Fig. 2.10. Logic network for F = $\overline{A.B + \overline{B.C}}$ in Example 2.9.

(b) The truth table development is shown in Fig. 2.11. Examination of the final column reveals that there is only *one* combination of input signals which produces a 1 for the function under consideration.

Therefore $\overline{A.B + \overline{B.C}} = \overline{A}.B.C$.

A	B	C	$A \cdot B$	$B \cdot C$	$\overline{B \cdot C}$	$A \cdot B + \overline{B \cdot C}$	$\overline{A \cdot B + \overline{B \cdot C}}$
0	0	0	0	0	1	1	0
0	0	1	0	0	1	1	0
0	1	0	0	0	1	1	0
0	1	1	0	1	0	0	1
1	0	0	0	0	1	1	0
1	0	1	0	0	1	1	0
1	1	0	1	0	1	1	0
1	1	1	1	1	0	1	0

Fig. 2.11. Truth table for Example 2.9.

The logic network for this simplified boolean expression is shown in Fig. 2.12.

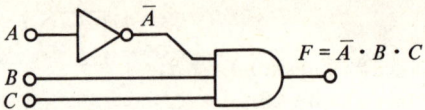

Fig. 2.12. Simplified logic network for Example 2.9.

EXAMPLE 2.10

Draw the logic network (using a *minimum* number of basic logic gates) which satisfies the truth table shown in Fig. 2.13.

To derive a logic network using a *minimum* number of gates we must first derive the boolean equation describing the function F, and then simplify the equation *before* representing it by logic gates.

A	B	C	F
0	0	0	0
0	0	1	0
0	1	0	1
0	1	1	0
1	0	0	0
1	0	1	1
1	1	0	1
1	1	1	1

Fig. 2.13. Truth table for Example 2.10.

The boolean equation describing the function F is obtained by writing down all the combinations of inputs for which a 1 is given for F as terms in the equation:

$$F = \bar{A}.B.\bar{C} + A.\bar{B}.C + A.B.\bar{C} + A.B.C$$

$$= \bar{A}.B.\bar{C} + A.B.\bar{C} + A.\bar{B}.C + A.B.C$$

$$= B.\bar{C}.(\bar{A} + A) + A.C(\bar{B} + B)$$

$$= B.\bar{C} + A.C$$

Thus, the simplest form for the boolean equation in this case is given by $F = A.C + B.\bar{C}$. The logic network is shown in Fig. 2.14.

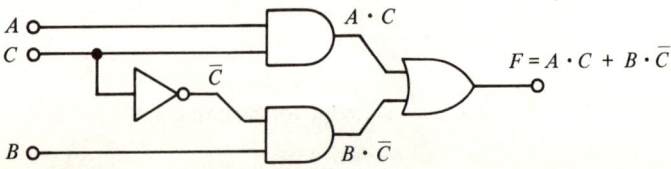

Fig. 2.14. Simplified logic network for Example 2.10.

EXERCISES TO CHAPTER 2

1. Simplify the boolean expression $(A + B) . 1$.
2. Simplify the boolean expression $A + B . 1$.
3. Simplify the boolean expression $(A + B) . 0$.
4. Simplify the boolean expression $A.B + A.B$.
5. Simplify the boolean expression $(A + B) . (A + B)$.
6. Simplify the boolean expression $(A + B) . (\overline{A + B})$.
7. Simplify the boolean expression $(A.\bar{B} + C) + (\overline{A.\bar{B} + C})$.
8. Use truth tables to prove that $A . (A + B) = A$ is true.
9. Use truth tables to prove that $(A + B) . (A + C) = A + B.C$ is true.
10. Simplify the boolean expression $A.B.C + A.\bar{B}.C + A.B.\bar{C}$, and draw a logic network (using basic gates) of the simplified function.
11. Use a truth table, simplify $A.B.\bar{C} + A.\bar{B}.C$, and draw a logic network (using basic gates) of the simplified function.
12. Draw the logic network which satisfies the truth table shown in Fig. 2.15, assuming that a minimum gate network is required which uses only AND, OR, and NOT gates.

A	B	C	F
0	0	0	0
0	0	1	0
0	1	0	0
0	1	1	0
1	0	0	0
1	0	1	1
1	1	0	1
1	1	1	0

Fig. 2.15. Truth table for Exercise 2.12.

3 Bistable elements

3.1 Introduction

Bistable elements have two *stable* operating states, in which the application of an input signal causes the device to change from one stable operating state to the other. Devices of this type are simple memories, since the state of the output at any instant can be used to deduce the state of the input at a previous instant.

It is unfortunate that in logic systems, bistable elements have been called (incorrectly) *flip-flops*. However, since the majority of manufacturers adopt the US Military Standard, which uses the name 'flip-flop' for bistable elements, it is necessary for students, technicians, and engineers to be aware of these facts.

3.2 The *S–R* bistable element

The logic symbol for the *S–R* bistable element is shown in Fig. 3.1, in which the two inputs S and R may be considered as SET and RESET. The outputs Q and $\overline{Q}$ imply that the logic states of the outputs are always opposite (or *complementary*) to each other.

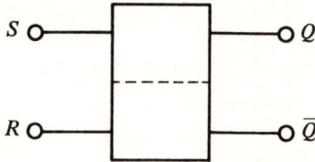

Fig. 3.1. Logic symbol of the *S–R* bistable element.

A logical 1 signal applied to the S input causes the Q output to be SET to logical 1 (and at the same time the $\overline{Q}$ output goes to logical 0). A logical 1 signal applied to the R input causes the Q output to be RESET to logical 0 (and at the same time the $\overline{Q}$ output goes to logical 1).

The *S–R* bistable can be constructed by interconnecting two cross-coupled 2-input NOR gates, as shown in Fig. 3.2. The simplified truth table, shown in Fig. 3.3(*a*), can easily be derived by assuming a logic state for the Q output and its complement for $\overline{Q}$, then applying the combinations of inputs to S and R and using the truth table for a NOR gate to derive the true output states. Note that the output of a NOR gate will only be 1 when the two inputs are at 0.

It must be noted that the changes which occur at the output *actually take place when the input signals change from 0 to 1*, i.e., if it is always assumed that $S = R = 0$ at the start (when power is applied), the output Q could be either 0 or

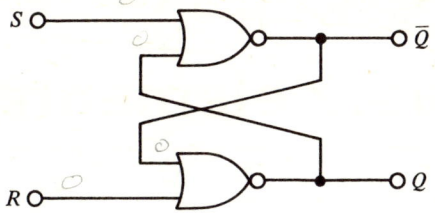

Fig. 3.2. *S–R* **bistable element — using cross-coupled NOR gates.**

S	R	Q	$\bar{Q}$
0	0	*	*
1	0	1	0
0	1	0	1
1	1	0	0

* Unpredictable at switch-on — depends
on previous input conditions — stable

(*a*) Simplified truth table

S	R	Q_{t-1}	Q_t	$\bar{Q}_t$
0	0	0	0	1
0	0	1	1	0
1	0	0	1	0
1	0	1	1	0
0	1	0	0	1
0	1	1	0	1
1	1	0	0	0
1	1	1	0	0

Q_{t-1} = Q state before application of stated signals
Q_t = Q state following application of stated signals

(*b*) Complete truth table

Fig. 3.3. Truth tables for NOR network *S–R* bistable element.

1; then if the signal applied to R is held at 0 while the signal applied to S is changed from 0 to 1, the output at Q will go to logic 1 (whatever its previous state). The signal applied to S can immediately be changed back to 0, since the 1 applied to S need only be applied momentarily for it to be *latched* in to Q. This means that the signals applied to S and R are both 0 again. Thus, the truth table for the cross-coupled NOR gate $S–R$ bistable is shown in Fig. 3.3(*b*), in which each set of conditions is applied by starting from and returning to $S = R = 0$.

Note: No useful information is produced at the outputs when simultaneous application of logical 1s is made to S and R (when the logic states produced at the outputs are both logical 0, which is contrary to the labelling of complementary outputs), therefore it is desirable in practical applications to avoid this condition.

The $S–R$ bistable element can also be constructed by interconnecting two cross-coupled 2-input NAND gates, as shown in Fig. 3.4. However, it should be noted that since NAND gates are *not the same* as NOR gates, the network shown in Fig. 3.4 is *not the same* as the network shown in Fig. 3.2. *Each* of these two networks performs the function of the $S–R$ bistable, but they *behave* differently. The simplified truth table, shown in Fig. 3.5(*a*), can easily be derived for the NAND network by assuming a logic state for the Q output and its complement

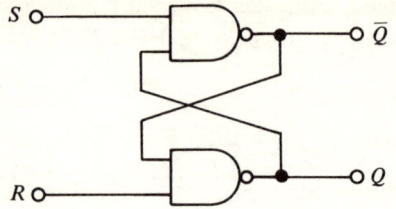

Fig. 3.4. *S-R* **bistable element, using cross-coupled NAND gates.**

S	R	Q	$\bar{Q}$
1	1	*	*
1	0	1	0
0	1	0	1
0	0	1	1

* Unpredictable at switch-on – depends
on previous input conditions – stable

(*a*) Simplified truth table

S	R	Q_{t-1}	Q_t	$\bar{Q}_t$
1	1	0	0	1
1	1	1	1	0
1	0	0	1	0
1	0	1	1	0
0	1	0	0	1
0	1	1	0	1
0	0	0	1	1
0	0	1	1	1

Q_{t-1} = Q state before application of stated signals
Q_t = Q state following application of stated signals

(*b*) Complete truth table

Fig. 3.5. Truth tables for NAND network *S–R* **bistable element.**

for $\bar{Q}$, then applying the combinations of inputs to S and R and using the truth table for a NAND gate to derive the true output states. Note that the output of a NAND gate will be logical 1 for all combinations of its inputs except when both inputs are at logical 1, when the output is at logical 0.

In the case of the NAND gate network, it must be noted that the changes which occur at the output *actually take place when the input signals change from logical 1 to logical 0*, that is, if it is always assumed that $S = R = 1$ at the start (when power is applied), the output Q could either be logical 0 or logical 1; then if the signal applied to S is held at logical 1 while the signal applied to R is changed from logical 1 to logical 0, the output at Q will go to logical 1 (whatever its previous state). The signal applied to R can immediately be changed back to logical 1, since the *change* applied to R need only be applied momentarily for it to be latched in to Q. This means that the signals applied to S and R are both at logical 1 again. Thus, the truth table for the cross-coupled NAND gate $S-R$ bistable is shown in Fig. 3.5(*b*), in which each set of conditions is applied by starting from and returning to $S = R = 1$.

Note: No useful information is produced at the outputs when simultaneous application of logical 0s is made to S and R (when the logical states produced at the outputs Q and $\bar{Q}$ are both logical 1, which is contrary to the labelling of complementary outputs), therefore it is desirable in practical applications to avoid this condition.

EXAMPLE 3.1

Draw the logic diagram of an arrangement of logic gates which is suitable to 'latch' a single short-duration positive-going input pulse.

A suitable arrangement of logic gates for this application is shown in Fig. 3.6, which is an *S-R* bistable using two cross-coupled NOR gates.

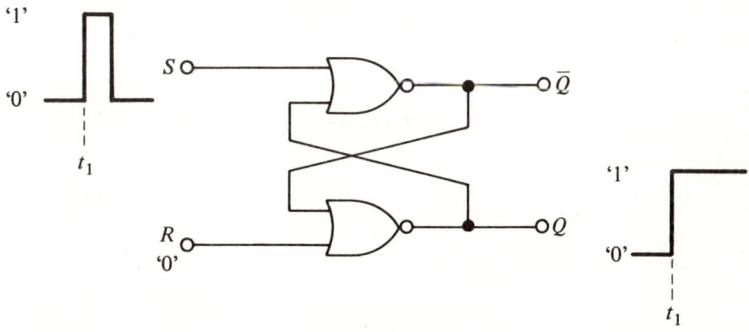

Fig. 3.6. Logic network for Example 3.1.

If it is assumed that the short-duration positive-going pulse is applied to the S input, and that the Q output is initially at logical 0, then the input and output waveforms will be as shown in Fig. 3.6.

3.3 The gated *S-R* bistable element

The logic symbol for the *gated S-R* bistable element is shown in Fig. 3.7(*a*). This is a modification of the simple *S-R* bistable by the addition of a third input called the *clock input*, CK, and can be realized by using the logic AND gates with the basic *S-R* bistable element as shown in Fig. 3.7(*b*). This device can now

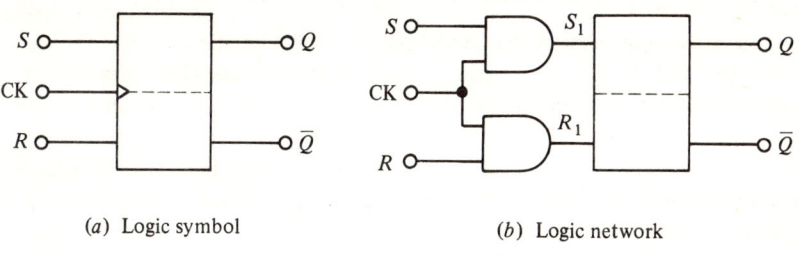

(*a*) Logic symbol   (*b*) Logic network

Fig. 3.7. The gated *S-R* bistable element.

be referred to as a *synchronous* device, since the logical 1 signals applied to the S and R inputs will only be passed through to S_1 and R_1 (the original bistable inputs) when the CK signal is at a logical 1 state. The clock signal thus enables the AND gates to transmit the signals applied at S and R through to S_1 and R_1.

The gated *S–R* bistable element is usually given additional inputs, PR (*preset*) and CLR (*clear*), as shown in Fig. 3.8. These inputs override the clock input, allowing the bistable to be *set* (PR) or *reset* (CLR) independent of the clock input.

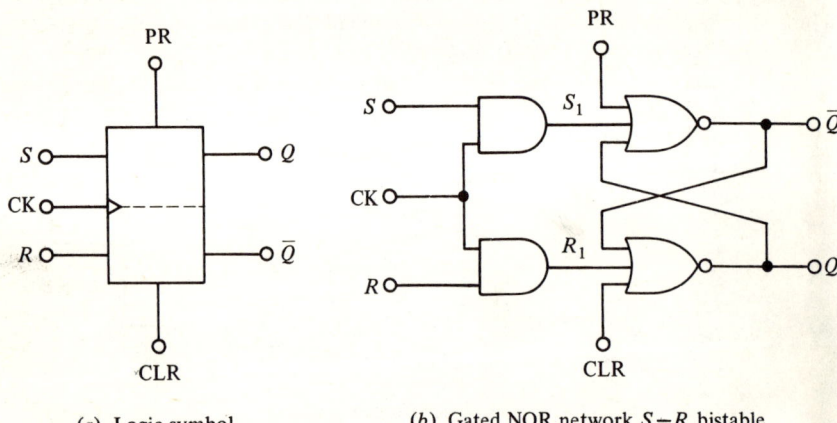

(a) Logic symbol   (b) Gated NOR network $S-R$ bistable

Fig. 3.8. Gated *S–R* bistable element

It should be noted that simultaneous application of logical 1 signals to S and R cause both outputs (Q and $\overline{Q}$) to go to logical 0 when the clock (CK) is at a logical 1 state. When the clock is at a logical 0 state, the bistable outputs return to *one* of their stable states.

The gated *S–R* bistable element can also be constructed using the two cross-coupled NAND gates to represent the basic *S–R* bistable.

EXAMPLE 3.2

Draw the logic diagram of an arrangement of logic gates which can be used to latch a short-duration positive-going pulse in synchronism with a clock pulse.

Since the requirement is to sense positive-going changes, then the NOR network *S–R* latch would be suitable. In order that the input pulse is *synchronized* with a clock input, then the gated *S–R* bistable network shown in Fig. 3.8(*b*) may be used.

If it is assumed that the Q output is initially at logical 0, that the R input is held at logical 0, that the clock input is applied to CK, and that the input pulse is applied to the S input, then as soon as the input pulse goes to logical 1 and the clock input CK is at logical 1, the output at Q goes to logical 1 and remains in that state.

3.4 The *D-type* bistable element

The *D-type* bistable element (or latch) was developed to overcome the limitation of the *S–R* bistable when simultaneous application of logical 1 states to the *S* and *R* inputs does not produce useful information at its outputs, and to provide a bistable element which has a single-signal input. The above-mentioned limitation is overcome by arranging that the *S* and *R* inputs are always complementary, as shown in Fig. 3.9.

A simplified truth table for the *D-type* bistable element is shown in Fig. 3.10, in which it is assumed that the stated conditions are *enabled* when the clock (CK) signal is at a logical 1 state, i.e., *leading-edge* triggered.

The logic symbol for the *D*-type latch shown in Fig. 3.9(*a*) indicates that the preset (PR) and clear (CLR) inputs are enabled by the application of logical 0 signals, *not* logical 1 signals as in the general case of *S–R* bistables.

Additionally it should be noted that, in practice, if no connection is made to the *D* input, a logical 1 level is generally assumed.

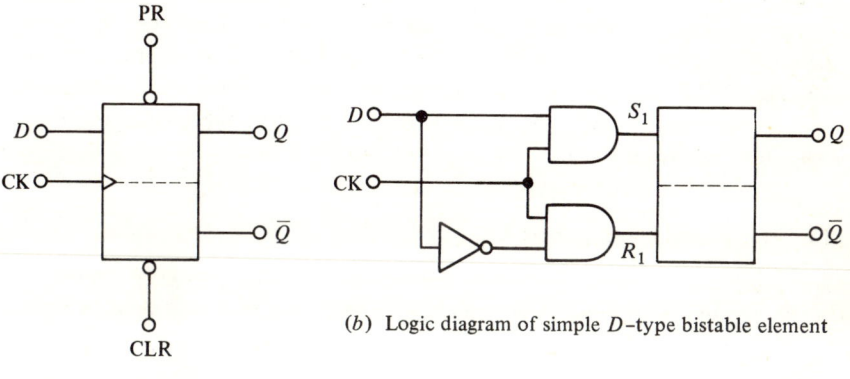

(*b*) Logic diagram of simple *D*–type bistable element

(*a*) Logic symbol

Fig. 3.9. The *D*-type bistable element.

D	Q_{t-1}	Q_t
0	0	0
0	1	0
1	0	1
1	1	1

Fig. 3.10. Truth table for the *D*-type bistable element.

EXAMPLE 3.3

Sketch the time-related waveforms for the input and output of the logic network shown in Fig. 3.11(*a*) assuming that TTL (transistor-transistor logic) devices are used.

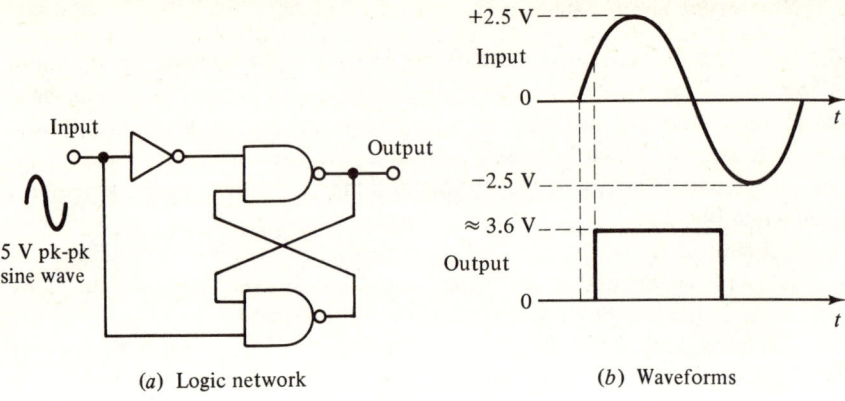

(a) Logic network   (b) Waveforms

Fig. 3.11. Logic network and waveforms for Example 3.3.

The logic network is identified as a simple (not gated) D-type bistable in which cross-coupled NAND gates are used. If it is assumed that the output is initially at logical 0 (which is the $\bar{Q}$ output) then the output will change to a logical 1 (approximately 3.6 V for TTL) when the input waveform is between 0.8 and 2.0 V. The logic network latches in this state until the input changes, so that when the input is between −0.8 and −2.0 V, the output changes to logical 0 (less than 0.4 V for TTL).

3.5 The 'master-slave' principle

The gated S–R bistable and the D-type bistable are edge-triggered systems, i.e., in the circuits considered above, the set or reset operation is initiated as soon as the clock signal changes from logical 0 to logical 1 on the *leading edge* of the positive pulse applied to the CK input. When bistable elements are connected in counting configurations this can cause timing problems; the *master-slave* technique was developed to improve this situation.

The master-slave technique uses *two* bistable elements in series, with their clock input signals arranged to be complementary, as shown in Fig. 3.12.

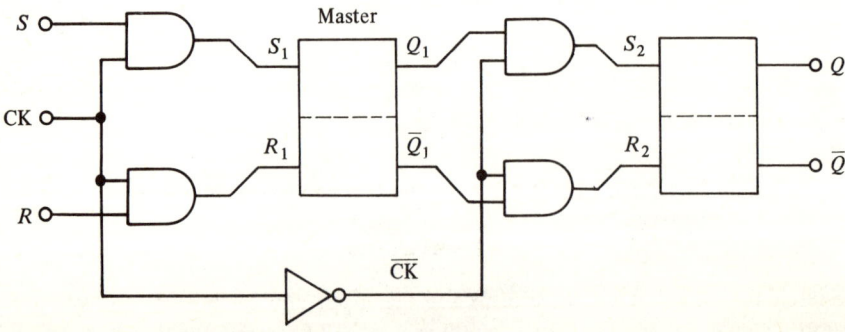

Fig. 3.12. Principle of the master-slave technique.

The *master* bistable element can only be set or reset when the clock input (CK) is logical 1, that is, on the leading edge of the pulse applied to CK. At the same time, the slave bistable element is inoperative (or *disabled*), since its clock input is at logical 0. When the clock input changes to logical 0, that is, on the trailing edge of the positive pulse applied to CK, the *slave* bistable is forced to change its state to that determined by the master output, since $\overline{CK}$ is now at logical 1. Thus, the system is primed on the leading edge of the clock pulse when the master is set (or reset). On the trailing edge of the clock pulse, the logic state of the master is transmitted to the output of the slave, so that the transition of an input signal to the master-slave bistable is completed when the clock pulse is on its trailing edge.

3.6 The *J–K* bistable element

When bistable elements are used as counting elements, some feedback is involved from output to input. With *edge-triggered* systems there is likely to be oscillation between one state and the other as long as the clock-input signal remains at logical 1. Hence the requirement for the master-slave system.

The *J–K* bistable element combines the capabilities of the gated *S–R* bistable with the master-slave technique, as shown in Fig. 3.13(*a*). The logic symbol for the *J–K* master-slave bistable element is shown in Fig. 3.13(*b*). Note that the symbolization indicates that this device is clocked (gated, or triggered) on the trailing edge of the clock input, and that the preset (PR) and clear (CLR) inputs are *enabled* by logical 0 signals.

A simplified truth table for the *J–K* bistable element is shown in Fig. 3.14, in which it is assumed that the stated conditions are enabled when the clock signal CK is changed from a logical 1 to logical 0, that is, on the *trailing edge* of the clock pulse.

It should be noted that simultaneous application of logical 1 signals to both *J* and *K* inputs cause the output *Q* to change its logical state on every clock pulse (trailing edge), and that the $\overline{Q}$ output is always in the complementary state to the *Q* output. Thus, if the *J* and *K* inputs are 'hard-wired' to a logical 1 level, then the *J–K* bistable element operates as a *toggle* bistable, as shown in Fig. 3.15(*a*), in which the output state at *Q* changes at the trailing edge of each clock pulse, as shown in Fig. 3.15(*b*). It should also be noted that the frequency of the output waveform is one-half of the frequency of the clock pulse, i.e., this is also a *divide-by-two* network or device.

EXAMPLE 3.4

Draw the time-related waveforms at the *Q* output *and* at the $\overline{Q}$ output of the arrangement, shown in Fig. 3.16(*a*), in response to the waveform applied. Assume that initially the *Q* state is at logical 0.

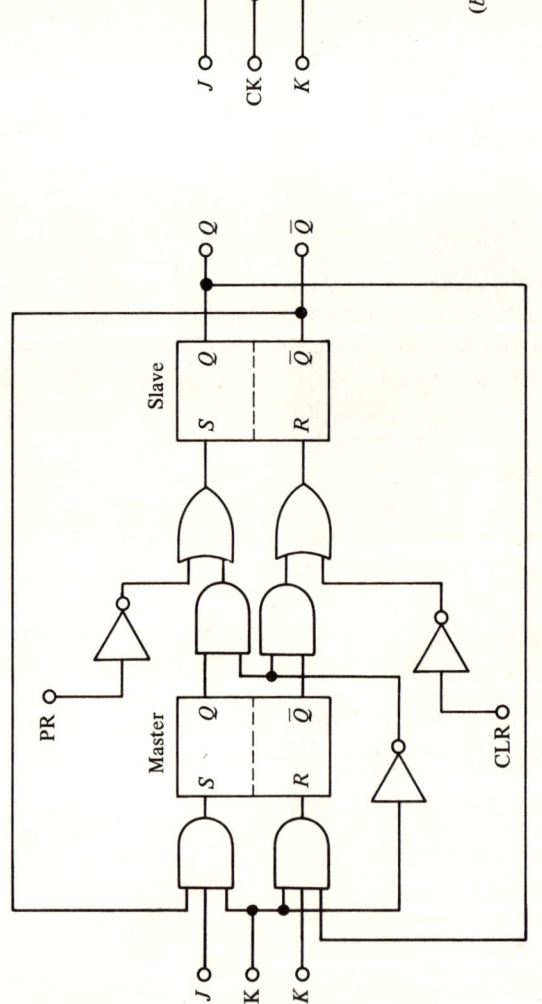

(b) Logic symbol

(a) Logic diagram

Fig. 3.13. The *J–K* bistable element.

J	K	Q_{t-1}	Q_t	$\bar{Q}_t$
0	0	0	0	1
0	0	1	1	0
1	0	0	1	0
1	0	1	1	0
0	1	0	0	1
0	1	1	0	1
1	1	0	1	0
1	1	1	0	1

Fig. 3.14. Simplified truth table for the J–K bistable element.

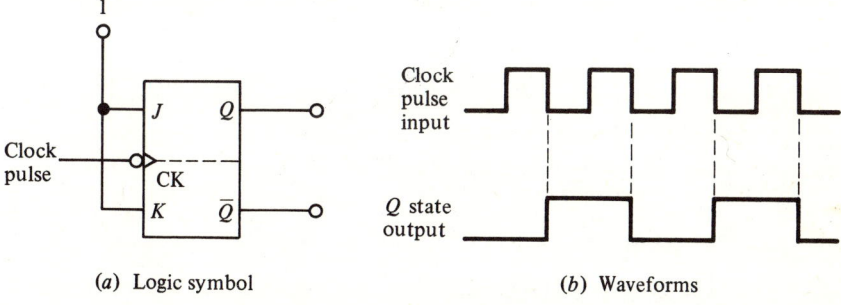

(a) Logic symbol   (b) Waveforms

Fig. 3.15. J–K as 'toggle' bistable.

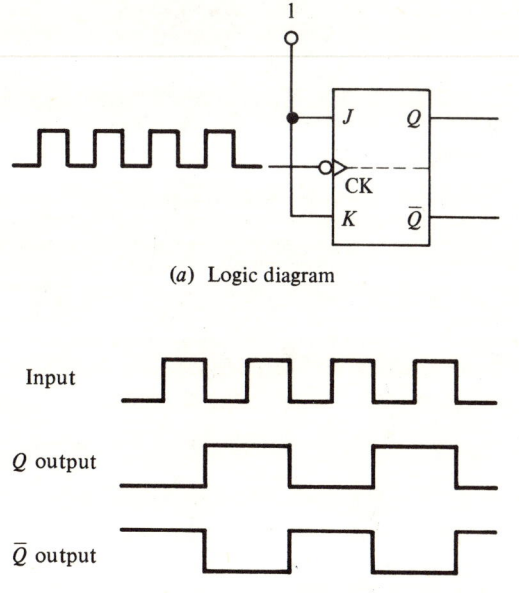

(a) Logic diagram

(b) Waveforms – solution

Fig. 3.16. Example 3.4.

EXERCISES TO CHAPTER 3

1. Draw the logic network for an *S-R* bistable using:

    (a) NAND gates only, and
    (b) NOR gates only.

2. Assuming that TTL logic gates are used, explain and derive the truth tables for *S-R* bistable elements using:

    (a) NAND gates only, and
    (b) NOR gates only.

3. Sketch a logic diagram to show how a *D*-type bistable element can be constructed from a simple *S-R* bistable using either NAND gates only *or* NOR gates only.

4. Explain the reasons for the need for a master-slave technique in bistable elements.

5. State the main advantage of a *J-K* bistable element when compared with a gated *S-R* bistable element.

6. Draw the logic symbol for a *J-K* bistable, and label all the terminals.

7. Using logic gates only, draw the logic diagram of a *J-K* bistable element.

8. Draw the logic diagram of a *D*-type bistable element, and describe its action.

9. Draw the logic diagram of a gated *S-R* bistable element, and explain its operation.

10. Sketch the logic network and label the connections of a circuit for which the output waveform is one-quarter of the frequency of the waveform applied at the input.

4 Counters

4.1 Introduction

Counters are generally made up from an assembly of bistable elements in which each bistable is essentially arranged as a *divide-by-two* or *modulo-2* counter. The operation of bistable elements as modulo-2 counters is shown in Fig. 4.1. The leading-edge-triggered D-type bistable is shown in Fig. 4.1(a), in which it is assumed that the Q output is initially at logical 0 (therefore the $\bar{Q}$ output is initially at logical 1, which is fed back to the D input). It can be clearly seen that the output waveform at Q is *half* the frequency of the pulses applied to the CK input. The master-slave (trailing-edge-triggered) J–K bistable is shown in Fig. 4.1(b), in which $J = K$ = logical 1, so that the Q output changes state at every trailing edge of the pulses applied to the CK input. The output waveform at Q is therefore *half* the frequency of the CK pulses.

4.2 Propagation delay and hazards

All logic devices take a finite time to operate, although the state changes and timing waveforms which are generally drawn seem to imply that the transitions occur instantaneously.

In logic networks, the change of a single variable from logical 0 to logical 1, or from logical 1 to logical 0, may produce a transient change in the output of the network when no change should occur. This problem is referred to as a *static hazard*. Consider the simple logic network shown in Fig. 4.2(a). In general, we assume that when $A = 1$ then $\bar{A} = 0$, and when $A = 0$ then $\bar{A} = 1$, which is certainly true for steady-state conditions. However, during transitionary periods, the condition $A = \bar{A} = 1$, or $A = \bar{A} = 0$ may occur. This is due to the time taken for the changed input signal to propagate through the NOT gate. The output conditions of the network will only be $A + \bar{A} = 0$ and $A.\bar{A} = 0$ at all times if the propagation delay is zero. But, in practice, the finite propagation delay causes the output $A + \bar{A}$ to fall to logical 0 for a short duration, and the output $A.\bar{A}$ rises to logical 1 for a short duration, as shown in Fig. 4.2(b). In most applications, this effect does not cause any real problem, but if these outputs supply circuits which count pulses, then these short-duration static hazards may be counted together with the required pulses.

Counters are made up from an assembly of bistable elements, as stated above, and if the output of the bistables changes too quickly, a spurious state (*glitch*) could *ripple* (or *race*) along the chain. This produces a *dynamic hazard*, or *race problem*, and is generally overcome during the design of the bistable elements. Leading-edge-triggered bistables are designed so that they take the logic level at their inputs just as the positive-going edge of the clock input occurs,

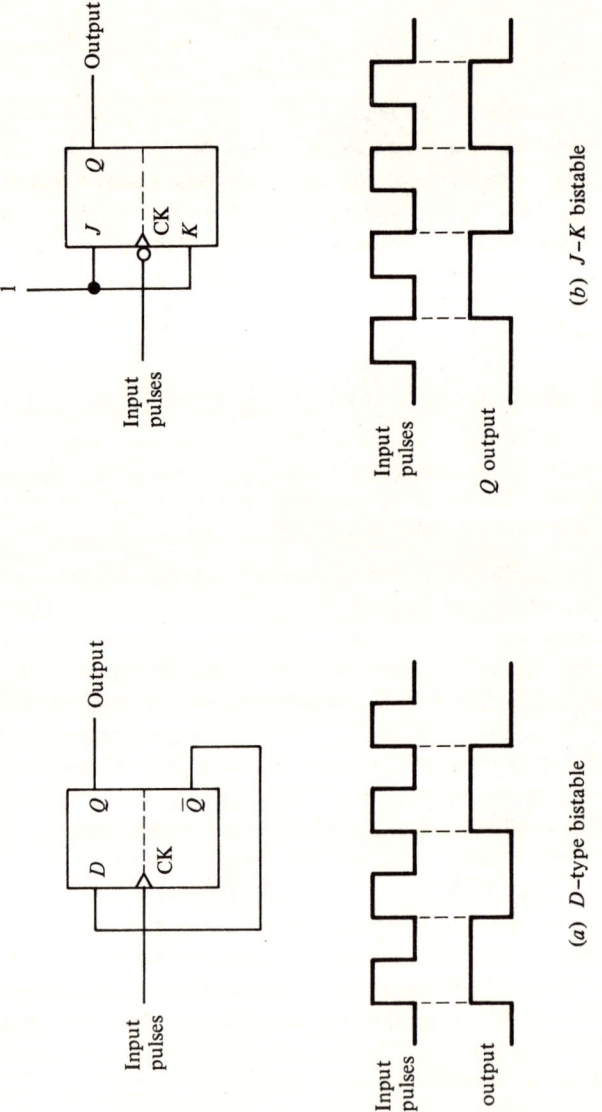

(a) D-type bistable

(b) J–K bistable

Fig. 4.1. Bistable elements as 'divide-by-two' or 'modulo 2' counters.

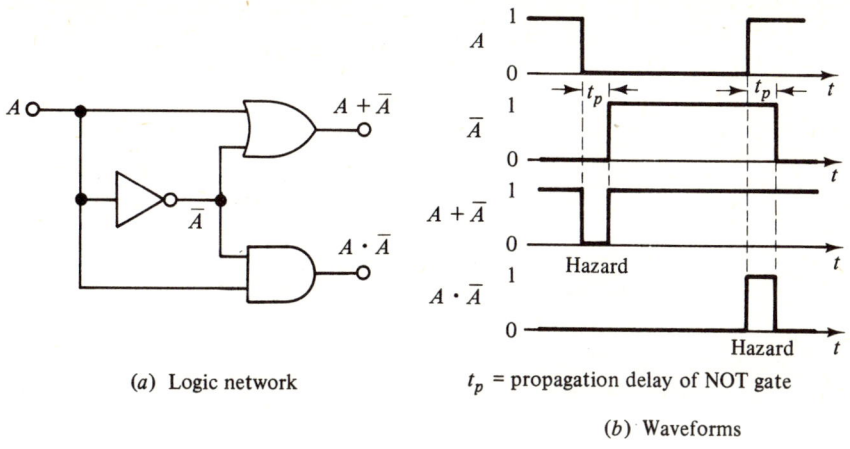

(a) Logic network t_p = propagation delay of NOT gate

(b) Waveforms

Fig. 4.2. Static hazards in logic networks.

and not slightly after or during the rise time of the clock. Master-slave or trailing-edge-triggered bistables are designed so that they take the logic level at their inputs just as the negative-going edge of the clock input occurs, and not slightly after or during the decay time of the clock. It is for this reason that practical clock pulse generators need clean edges, i.e., fast rise and fall times.

4.3 Asynchronous (ripple-through) counters

An *asynchronous counter* is a sequential logic system in which the pulses to be counted are applied at one end of the counter, i.e., in *serial* form, one pulse following the other, and the process of adding each pulse must be completed before the *carry bit* is propagated to the following stage. This next stage must then add the carry bit to the number in that stage, i.e., the carry bit appears to *ripple through* the length of the counter until the count is complete.

A 4-bit asynchronous binary counter using D-type bistable elements is shown in Fig. 4.3.

If it is assumed that the Q output states are initially all at logical 0, and the pulses to be counted are applied to the CK input of the first bistable element, then the timing waveforms (which indicate the time-related logical states throughout the counter) will be as shown in Fig. 4.4. This sequence will be repeated as successive pulses are applied to the CK input. It should be noted that initially, since all Q outputs are assumed to be at logical 0, all the $\bar{Q}$ outputs will initially be at logical 1, and the $\bar{Q}$ outputs are fed back to the D inputs in each stage of the counter. The 'count' can be determined at any instant from the timing waveforms in Fig. 4.4, as shown for a count of 7_{10}, that is, 0111_2.

The state of the count at any instant can also be determined with the aid of a truth table, as shown in Fig. 4.5. When the count is read from the Q output states (in parallel) of the bistable elements in Fig. 4.3, the sequential output states are that of an *up-counter*, which can be seen in Figs 4.4 and 4.5.

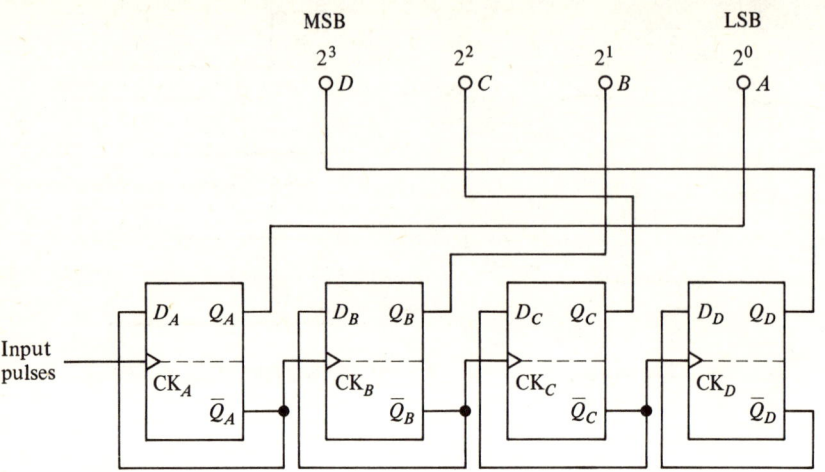

Fig. 4.3. **4-bit asynchronous binary counter — using** D**-type bistables.**

However, if the counter outputs are taken from the $\overline{Q}$ outputs of the bi-stable elements in Fig. 4.3, the sequential output states correspond to that of a *down-counter*, which can also be seen from Figs 4.4 and 4.5.

An alternative asynchronous binary counter can be constructed using J–K bistable elements, as shown in Fig. 4.6. Since the J–K bistable elements are trailing-edge-triggered, and all the J and K inputs are connected to logical 1, each Q output state changes on the trailing edge of the signal applied to the bistable element's CK input, as shown in the time-related waveforms in Fig. 4.7. Examination of these waveforms shows that the active pulse is effectively a negative-going pulse. The sequence shown is repeated as successive pulses are applied to the CK input.

The truth table shown in Fig. 4.5 also applies to the 4-bit binary counter shown in Fig. 4.6, so that if the counter output is read from the Q output states, it is an up-counter, and if read from the $\overline{Q}$ output states, it is a down-counter. Logic networks could be devised to be used with either of these counters to convert them into *reversible* counters, i.e., to use a single 'control' signal to select whether an up-counter or down-counter function will be executed.

4.4 Ripple-through delay

In practice, each bistable element has a finite propagation delay, i.e., a time delay between the input signal change and the corresponding output state change occurring. The effect of the actual propagation delay throughout the counter shown in Fig. 4.6, during the transition from 0011_2 to 0100_2 (3_{10} to 4_{10}), is shown in Fig. 4.8, in which it is assumed that the propagation delay of each bistable element is t_p.

The change from 0011_2 to 0100_2 is therefore accomplished in $3t_p$ seconds, with two erroneous transitionary states occurring during the total change.

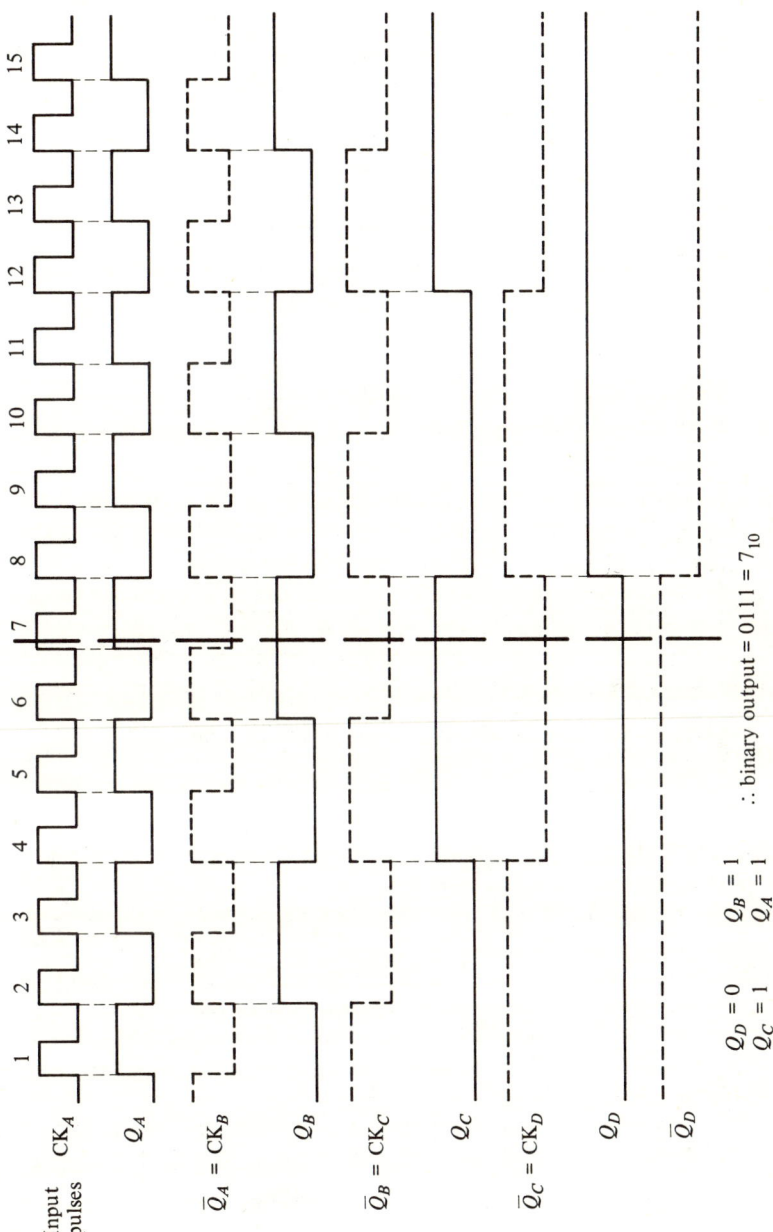

Fig. 4.4. Time-related waveforms for the asynchronous counter shown in Fig. 4.3.

CK	Q_D	Q_C	Q_B	Q_A	$\overline{Q}_D$	$\overline{Q}_C$	$\overline{Q}_B$	$\overline{Q}_A$
0	0	0	0	0	1	1	1	1
1	0	0	0	1	1	1	1	0
2	0	0	1	0	1	1	0	1
3	0	0	1	1	1	1	0	0
4	0	1	0	0	1	0	1	1
5	0	1	0	1	1	0	1	0
6	0	1	1	0	1	0	0	1
7	0	1	1	1	1	0	0	0
8	1	0	0	0	0	1	1	1
9	1	0	0	1	0	1	1	0
10	1	0	1	0	0	1	0	1
11	1	0	1	1	0	1	0	0
12	1	1	0	0	0	0	1	1
13	1	1	0	1	0	0	1	0
14	1	1	1	0	0	0	0	1
15	1	1	1	1	0	0	0	0

Fig. 4.5. Truth table for the asynchronous binary counter of Fig. 4.3.

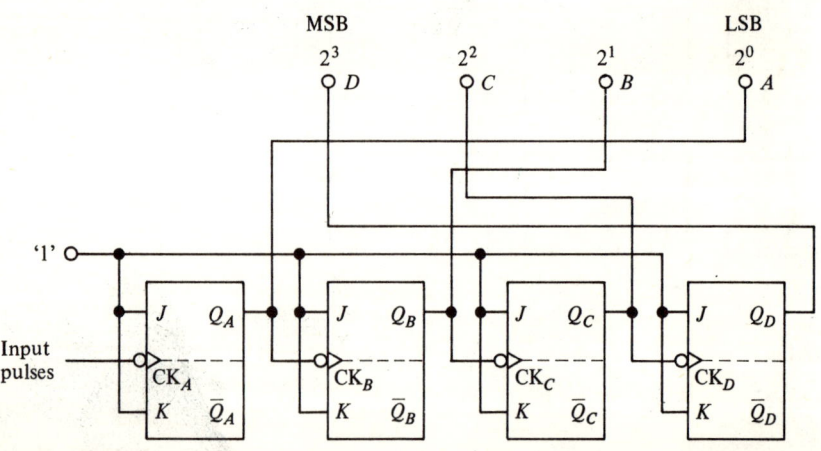

Fig. 4.6. Asynchronous binary counter – using J–K type bistables.

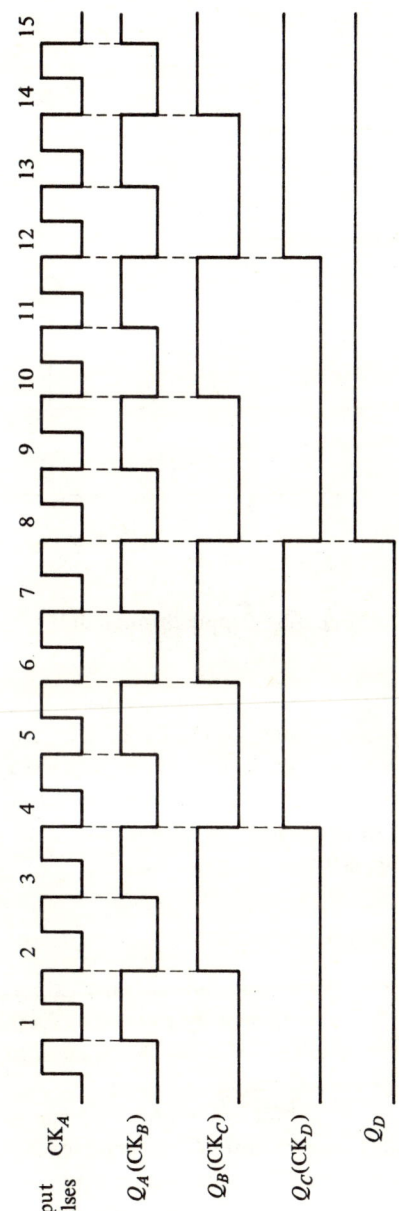

Fig. 4.7. Time-related waveforms for the asynchronous counter shown in Fig. 4.6.

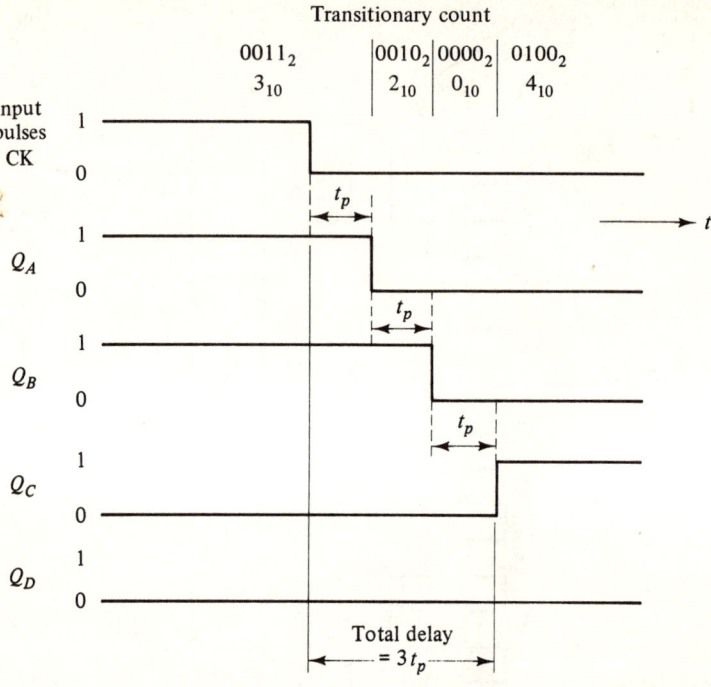

Fig. 4.8. Ripple-through delay.

Normally t_p is very small, being typically a few nanoseconds, and in many applications the ripple-through delay may be neglected, when the waveforms shown in Figs 4.4 and 4.7 are valid. However, in very high speed systems, the ripple-through delay may become significant, and dynamic hazards may more easily have a noticeable effect.

4.5 Synchronous counters

In *synchronous counters*, the counting sequence is controlled by means of a clock pulse (CK) applied simultaneously to all the bistable elements, causing all the *changes* of all the bistables to occur in synchronism. This eliminates the propagation delay experienced in ripple-through counters.

The logic network of a 4-bit synchronous binary counter is shown in Fig. 4.9. The Q output state of each bistable element (except the first) takes on the logic state applied to its J and K input at the next negative-going (trailing) edge of the CK input pulse. The first (least significant bit—LSB) bistable *changes* its logical output Q at *every* trailing edge of the CK input pulse. The time-related waveforms for this counter are very similar to those shown for the asynchronous counter in Fig. 4.7. The main difference is in the operation of the counter; in the synchronous counter the change of state of each bistable element is dependent

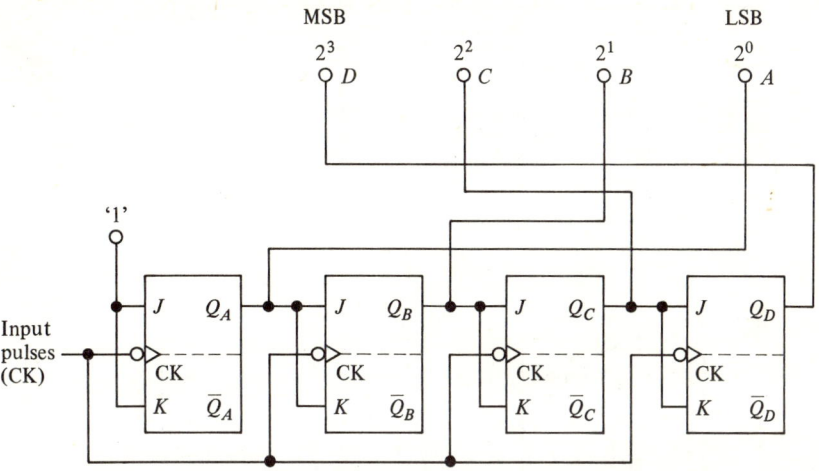

Fig. 4.9. Synchronous 4-bit binary counter.

upon the logic level applied to its J and K input and occurs on the trailing edge of the CK input, which is applied simultaneously to all bistable elements in the counter.

The truth table for the synchronous binary counter is also the same as that for the asynchronous counter as shown in Fig. 4.5. It can therefore be seen that by reading the parallel output from the Q outputs, an up-counter is produced, and by reading the $\overline{Q}$ outputs a down-counter is produced.

4.6 Divide-by-sixteen counter

An examination of both the asynchronous counter shown in Fig. 4.6 and the *synchronous* counter shown in Fig. 4.9, *together with* the timing waveforms shown in Fig. 4.7 reveals that the Q_D output state *only changes from logical 1 to logical 0 after 16 pulses have been applied to the CK input* of the counters. Since the bistable elements used in these counters are triggered by trailing edges, i.e., a logical 1 to logical 0 transition, the *single Q_D output* could be used to trigger another logic network or counter stage at *one-sixteenth* of the frequency of the CK input pulse, as shown in Fig. 4.10, provided that the driven network is activated (or enabled) by a trailing edge.

4.7 Modification of counting period

The counters described so far are binary counters, i.e., their counting periods are limited to numbers which are all powers of two. The methods which may be used to curtail the counting period to any other number generally depend upon the availability of a *clear* (CLR) facility in the bistable elements.

The clear (CLR) terminals on both D-type and J-K TTL bistable elements are generally enabled by a logical 0 level—thus, this terminal is disabled while a

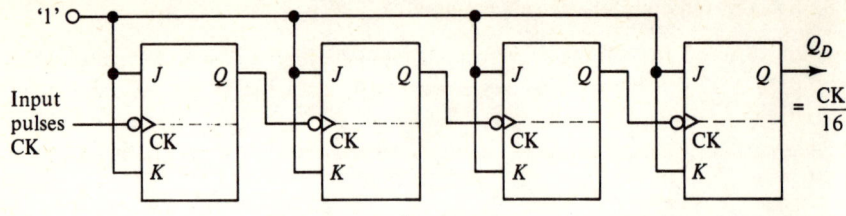

(a) Asynchronous divide-by-sixteen counter

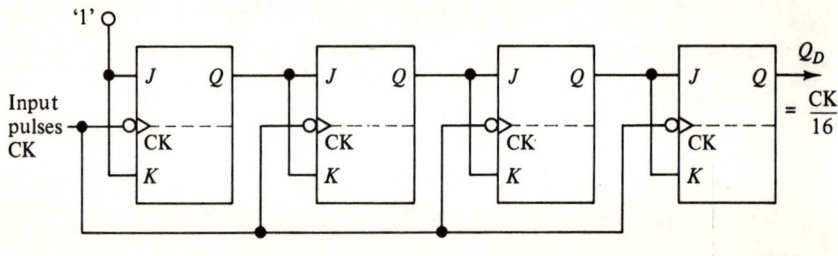

(b) Synchronous divide-by-sixteen counter

(c) Waveforms of divide-by-sixteen counter

Fig. 4.10. Divide-by-sixteen counters.

logical 1 is applied to it. The required count state must therefore be *detected* with the aid of logic gates to produce a logical 0 which must be connected to all the clear (CLR) terminals of all the bistable elements. During the normal counting conditions, the logical level applied to the CLR inputs must be at logical 1 level. The NAND logic gate is therefore suitable for this purpose, since its output will only be a logical 0 when all its inputs are at logical 1; the output is logical 1 for all other combinations of input states.

Note: It is actually necessary to connect the output of the logic network (*decoding* network) to the CLR input of only those bistable elements whose outputs are at logical 1.

4.8 Scale-of-ten counter—decade counter

Examination of the timing waveforms shown in Fig. 4.7 and the truth table shown in Fig. 4.5 shows that after ten (10) pulses the Q output states are: $Q_D = 1$, $Q_C = 0$, $Q_B = 1$, and $Q_A = 0$, that is, $1010_2 = 10_{10}$. However, for the *decade* counter, instead of adopting this state we require *all* the Q output states

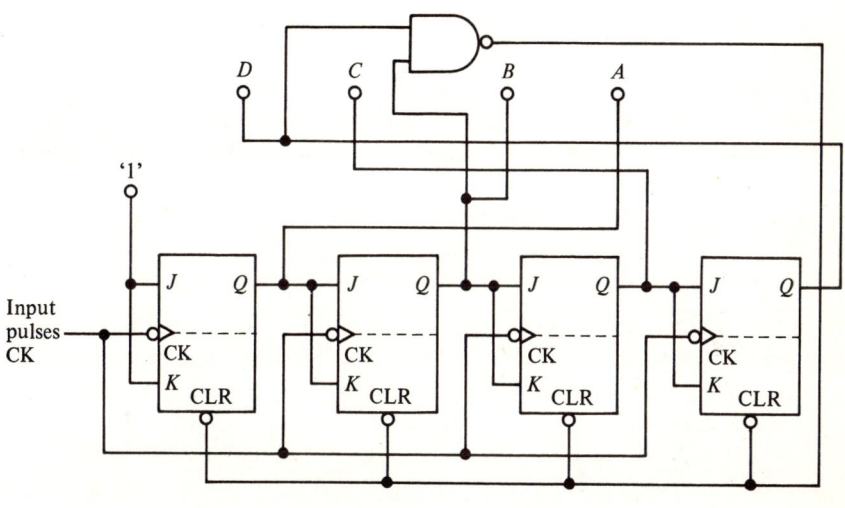

(*a*) Logic diagram of decade counter

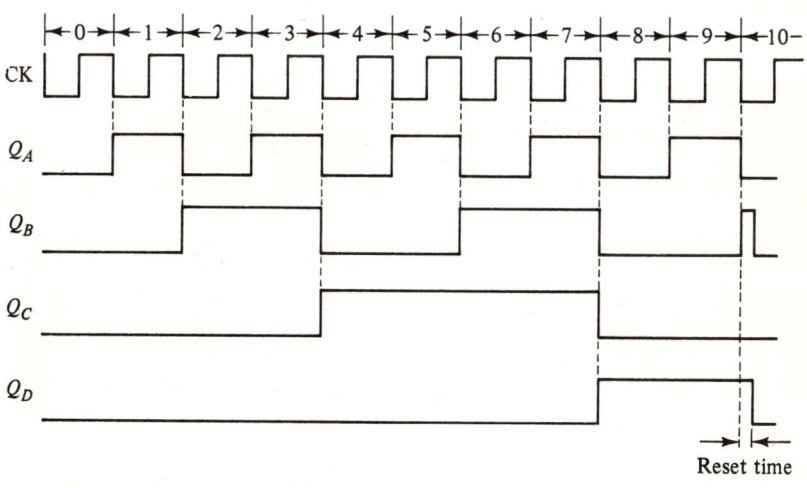

(*b*) Timing waveforms

Fig. 4.11. Logic diagram and waveforms for decade counter.

to be at logical 0. If a 2-input NAND gate is used with its inputs connected to Q_D and Q_B, the output of the NAND gate will be logical 0 for the count condition stated above. If the output of the NAND gate is connected to the clear (CLR) terminals of all the bistable elements as shown in Fig. 4.11(a), then the count will return to zero, that is, 0000_2, provided that the propagation delay of the clear inputs is short, as shown in Fig. 4.11(b). The reset time is typically 50 ns for TTL devices.

4.9   Curtailing count to any number

The principle described above may be applied to any of the binary counters considered so far in order to curtail the count to any number less than 2^n, where n = number of binary stages, provided that it is remembered that it is the state *following* the highest number to be counted that must be detected.

EXAMPLE 4.1

For a *scale-of-six* counter, the count sequence is 0, 1, 2, 3, 4, 5 which is repeated as further clock pulses are applied. In this case, the state to be detected is 6_{10}, which corresponds to $Q_C = 1$, $Q_B = 1$, and $Q_A = 0$, that is, 110_2. Therefore Q_B and Q_C must be connected to the inputs of a 2-input NAND gate, and the output of the NAND gate is connected to the CLR terminals of the bistable elements, as shown in Fig. 4.12, in which the 3-bit asynchronous counter is curtailed to a scale-of-six counter.

　　Further examples of the logic states to be detected—with NAND gates—for a range of counters are given below:

1.　Scale-of-three counter—detect states Q_B and Q_A.
2.　Scale-of-four counter—detect state of Q_C only, and use a NOT gate.
3.　Scale-of-five counter—detect states Q_C and Q_A.
4.　Scale-of-seven counter—detect states Q_C, Q_B, and Q_A.
5.　Scale-of-eight counter—detect state of Q_D, and use a NOT gate.
6.　Scale-of-nine counter—detect states Q_D and Q_A.

4.10   8421 BCD counter

Although the binary number system is the simplest and most suitable for digital computers, it is not compatible with human activity and human interpretation— the denary number system is the most familiar system from the human point of view. Unfortunately, there is no convenient 'chip' or system which we can use to convert a binary number into a denary number and vice versa.

　　The most convenient method by which the conversion can be achieved— using hardware—is by converting the binary count into an 8421 BCD count, and then decoding the BCD into denary numbers (Sec. 4.13). The 8421 BCD uses a 4-bit binary group to represent each denary digit—where the 4-bit binary groups correspond to the straight binary 4-bit groups for the denary digits 0 through 9.

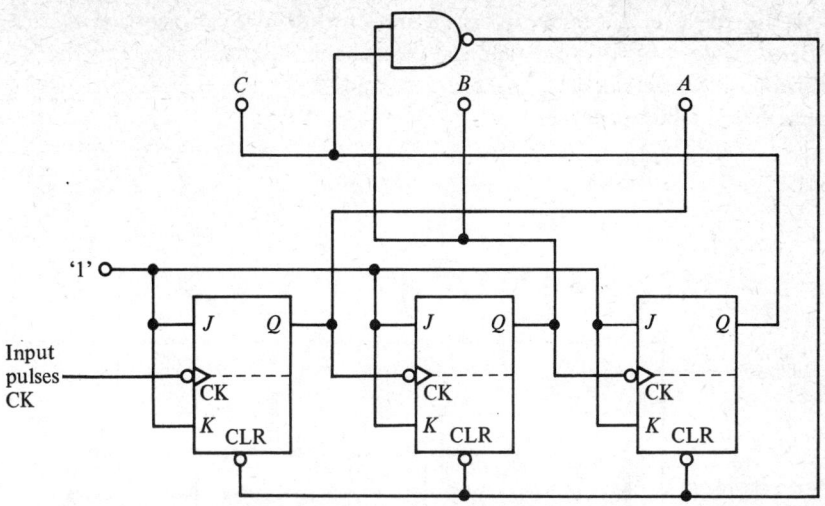

Fig. 4.12. Scale-of-six asynchronous counter.

Thus, the 8421 BCD counter is a 4-bit binary counter whose count is curtailed to a scale-of-ten counter. The logic diagram of an asynchronous 8421 BCD counter is shown in Fig. 4.13, in which a single logic gate is used.

4.11   TTL decade counter—SN 7490

The SN 7490 TTL decade counter is a synchronous counter which effectively produces an 8421 BCD output. This counter is *actually* made up from two

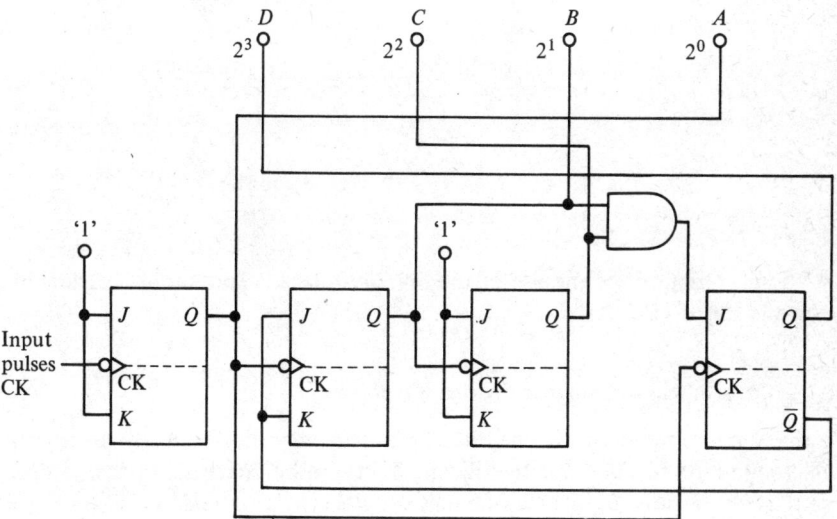

Fig. 4.13. Asynchronous 8421 BCD counter.

separate stages—a divide-by-two stage and a divide-by-five stage—and therefore offers a wide range of varied applications. The arrangement shown in Fig. 4.14 illustrates the *pin-outs* of this package, and the connections for its use as a decade (8421 BCD) counter.

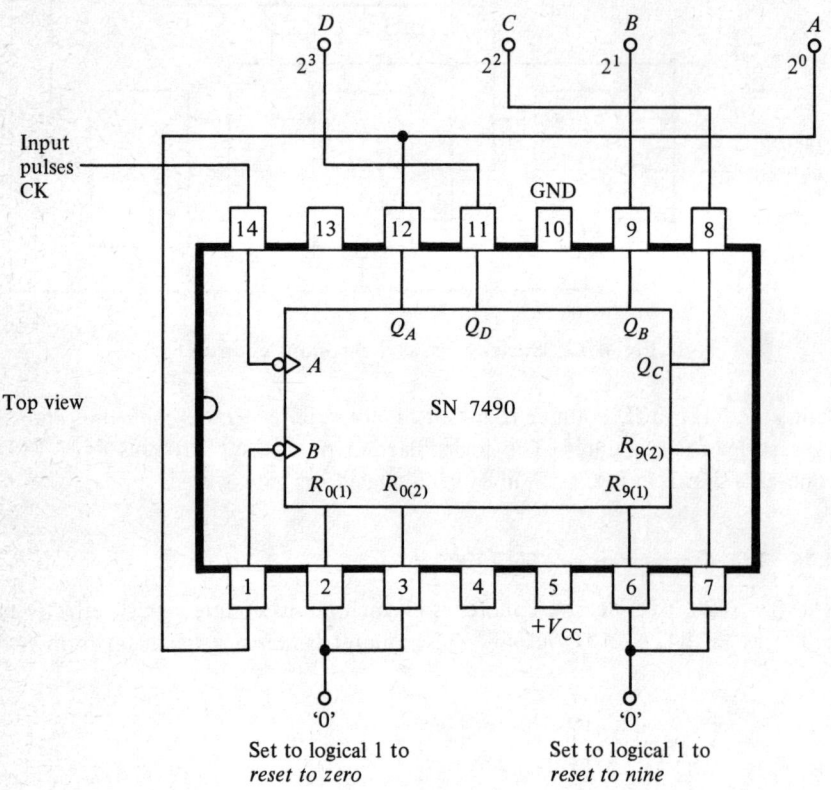

Fig. 4.14. TTL SN 7490 decade counter.

Some variations of the use of the SN 7490 decade counter as a divider are shown in Fig. 4.15.

4.12 Squaring-up asymmetric divided waveforms

In many microprocessor systems the clock frequency may be many megahertz, which could be too high for the digital circuits to be driven. A further requirement of the digital circuits may be that the clock pulse should be a symmetrical waveform. If counter circuits are used to divide the high-frequency source, then the waveform produced is asymmetric. This waveform may be converted into a

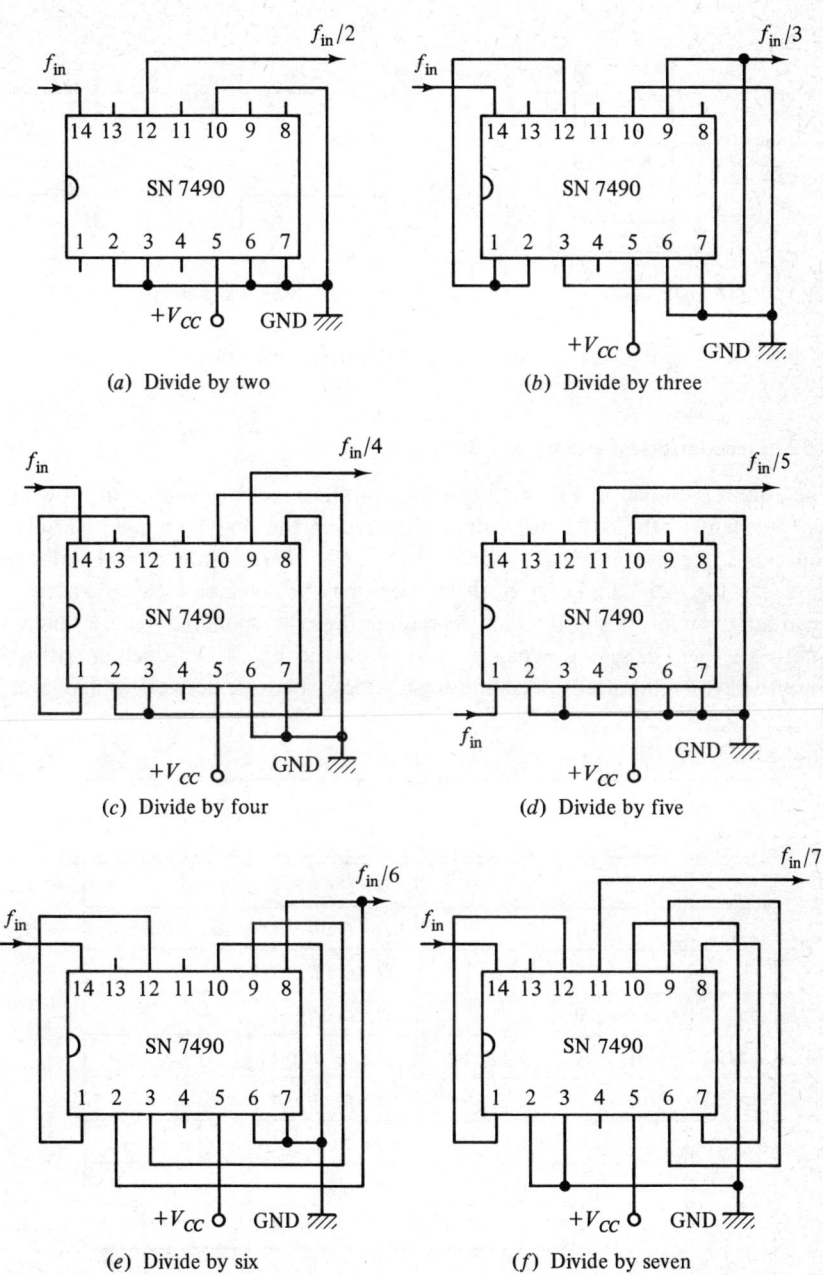

Fig. 4.15. Divider applications of SN 7490.

square wave by using the bistable arrangement shown in Fig. 4.16(*a*), with typical waveforms being shown in Fig. 4.16(*b*). It should be noted that this bistable element also introduces a further divide-by-two factor.

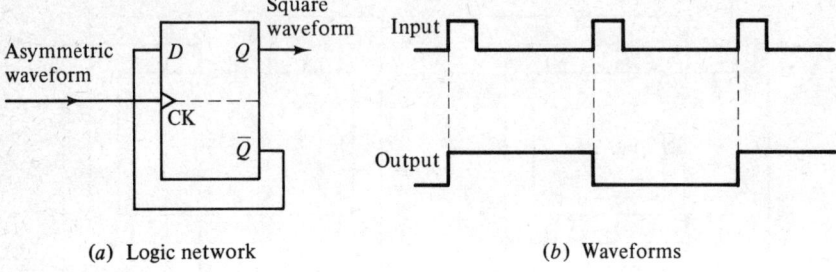

(*a*) Logic network   (*b*) Waveforms

Fig. 4.16. **Squaring-up asymmetric waveform.**

4.13 Decoder/driver and denary display

The counters shown in Figs 4.13 and 4.14 produce a binary count output which is equivalent to the 8421 BCD, since they repeat the first 10 sequential binary codes as successive pulses are applied to their CK inputs. These output states are shown in Fig. 4.17, and each of these states may be detected by a logic network in order to supply a signal which represents the corresponding denary number. The basic logic decoding signals are also shown in Fig. 4.17, together with the simplified (or minimized) decoding logic states. The logic network which gives a

Denary value	Q_D	Q_C	Q_B	Q_A	Decoding logic	Minimized decoding logic
0	0	0	0	0	$\bar{D} \cdot \bar{C} \cdot \bar{B} \cdot \bar{A}$	$\bar{D} \cdot \bar{C} \cdot \bar{B} \cdot \bar{A}$
1	0	0	0	1	$\bar{D} \cdot \bar{C} \cdot \bar{B} \cdot A$	$\bar{D} \cdot \bar{C} \cdot \bar{B} \cdot A$
2	0	0	1	0	$\bar{D} \cdot \bar{C} \cdot B \cdot \bar{A}$	$\bar{C} \cdot B \cdot \bar{A}$
3	0	0	1	1	$\bar{D} \cdot \bar{C} \cdot B \cdot A$	$\bar{C} \cdot B \cdot A$
4	0	1	0	0	$\bar{D} \cdot C \cdot \bar{B} \cdot \bar{A}$	$C \cdot \bar{B} \cdot \bar{A}$
5	0	1	0	1	$\bar{D} \cdot C \cdot \bar{B} \cdot A$	$C \cdot \bar{B} \cdot A$
6	0	1	1	0	$\bar{D} \cdot C \cdot B \cdot \bar{A}$	$C \cdot B \cdot \bar{A}$
7	0	1	1	1	$\bar{D} \cdot C \cdot B \cdot A$	$C \cdot B \cdot A$
8	1	0	0	0	$D \cdot \bar{C} \cdot \bar{B} \cdot \bar{A}$	$D \cdot \bar{A}$
9	1	0	0	1	$D \cdot \bar{C} \cdot \bar{B} \cdot A$	$D \cdot A$

Fig. 4.17. **BCD count sequence, and decoding logic.**

separate output corresponding to each denary number is referred to as an *8421 BCD-to-decimal converter* as shown in Fig. 4.18(*a*). A TTL BCD-to-decimal converter—the SN 7442A—is shown in Fig. 4.18(*b*).

It is commonly required to display the output of counters in a denary form. This can be accomplished by using 7-segment LED displays, of which there are two types, known as *common anode* and *common cathode*. In the common-anode type, all the segment anodes are connected together and then to $+V_{CC}$, and all the individual segments are illuminated by connecting that segment cathode to a logical 0 level—generally through a limiting resistor (typically 240 Ω to 2 kΩ). In the common-cathode type, all the segment cathodes are connected together and then to logical 0 (GND), and all the individual segments are illuminated by connecting that segment anode to a logical 1 level ($+V_{CC}$)— usually through a limiting resistor (typically 240 Ω to 2 kΩ). A common-anode 7-segment LED display is shown in Fig. 4.19(*a*), and a suitable *TTL BCD-to-7-segment decoder/driver*—the SN 7447A—is shown in Fig. 4.19(*b*), which is suitable for driving a common-anode 7-segment display.

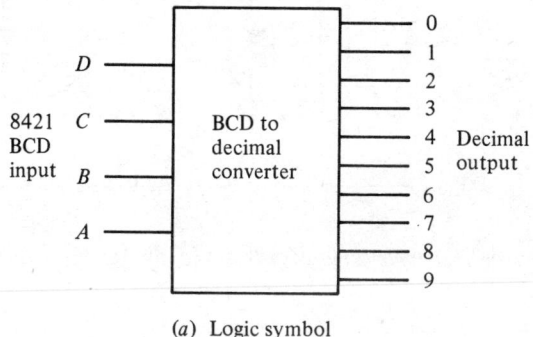

(*a*) Logic symbol

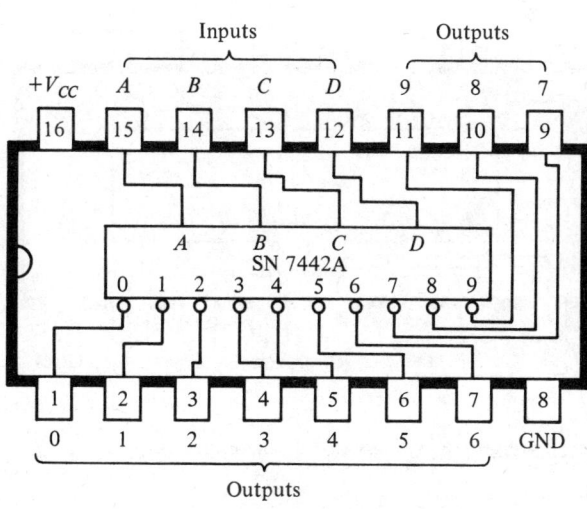

(*b*) Pin outs of SN7442A

Fig. 4.18. BCD-to-decimal decoder.

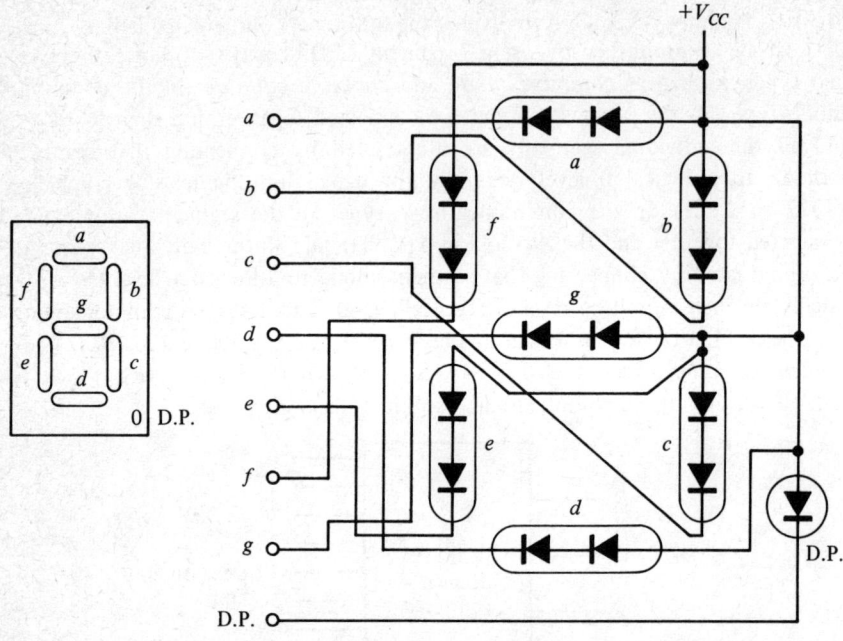

(a) Common-anode 7-segment LED display

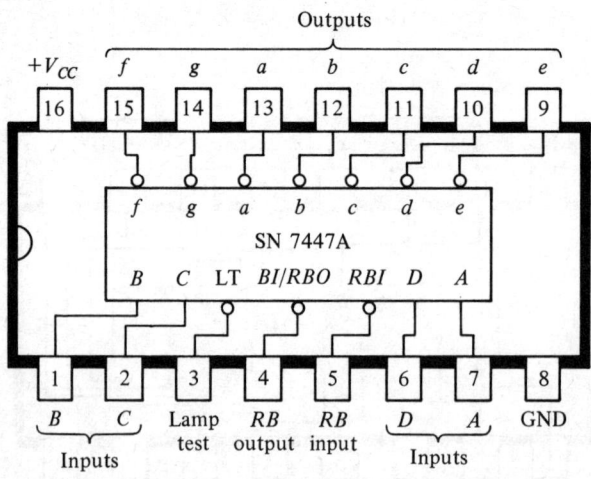

(b) SN7447A – BCD to 7-segment decoder/driver

Fig. 4.19. 7-segment LED display and BCD-to-7-segment decoder.

4.14 CMOS ripple-through counters

As previously stated, ripple-through counters use the output of any stage as the input clock of the next stage and therefore suffer from the problem of propagation delay from input to output, particularly for counters having large numbers of stages. CMOS devices are slower than TTL, so that the propagation delay in CMOS counters is longer than in TTL counters.

The availability of counters having many stages is much more common with CMOS devices than with TTL devices, with counters up to 21 stages being currently available in CMOS devices. Counters having large numbers of stages are particularly useful for large-scale frequency dividers. However, due to the limitation on the number of pin connections, not all of the stage outputs are available from CMOS counters.

The speed of operation of CMOS devices is generally dependent on the level of supply voltage used. CMOS ripple-through counters will operate successfully using typical clock-pulse frequencies of 5 MHz at 5 V, 15 MHz at 10 V, and 20 MHz at 15 V supply. It is always advisable to consult manufacturer's data for actual figures for a particular device. Virtually all CMOS ripple counters are clocked on the trailing edge of the clock input pulse.

Typical CMOS ripple counters include 7-stage (4024), 12-stage (4040), 14-stage (4020 and 4060), and 21-stage (4045).

The pin connections of the CMOS 4020 14-stage ripple counter are shown in Fig. 4.20. It should be noted that no outputs exist for the second and third stages. This counter can provide outputs ranging from the first stage which divides the clock input by 2^1 (2) to the 14th stage which divides the clock input by 2^{14} (16384).

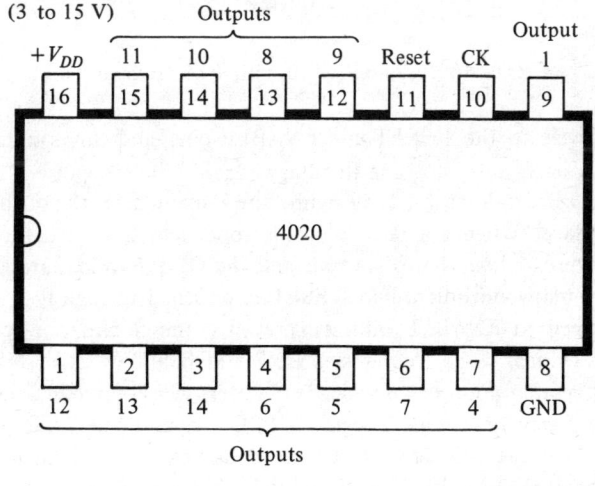

Top view

Fig. 4.20. CMOS 4020 14-stage ripple counter.

4.15 CMOS synchronous counters

The internal counter stages of synchronous counters are driven from single clocked inputs and all the outputs change simultaneously. CMOS synchronous counters are generally classed as follows:

(a) *Decoded 1-of-N* counters are mainly used in such applications as digital filters and waveform generators. Typical devices are the 4017 which is a decoded 1-of-10, and the 4022 which is a decoded 1-of-8 device.

    The 4017 is a fully synchronous decade (or divide-by-10) counter, which may also be used to obtain a 1-of-10 decoded output. The pin-out diagram is shown in Fig. 4.21. Normal operation is effected by applying logical 0

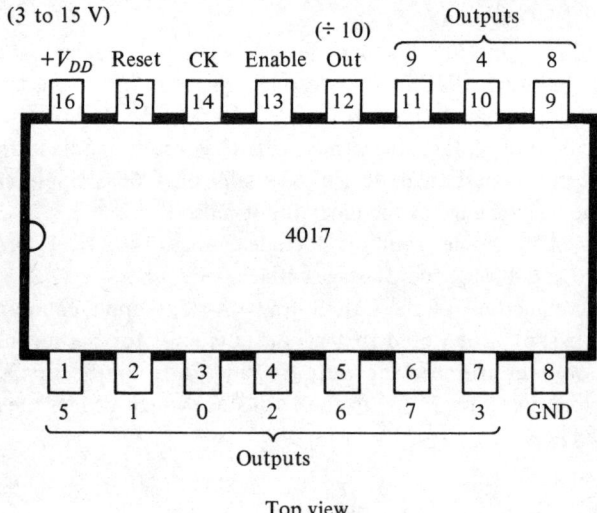

Fig. 4.21. CMOS 4017 divide-by-10 with 1-of-10 decoded outputs.

(GND) levels to the RESET and ENABLE pins, and the counter advances by one at each positive-going (leading) edge of the CK pulse. The decoded output goes to a logical 1 level, while the remainder of the outputs are at a logical 0 level. When a logical 1 level is applied to the reset, the counter is reset to zero—when the '0' output and the OUT terminal are at logical 1. Counting may continue when RESET is returned to logical 0. A logical 1 level applied to ENABLE inhibits (prevents) the clock from operating the counter. Typical clock frequencies are 2.5 MHz at 5 V and 5 MHz at 10 V.

(b) *Display-decoded* counters are decade-counters which provide a decoded output suitable for producing 7-segment LED display data. Typical devices are the 4026 and the 4033, each of which require external drivers to supply the high-current LED displays.

    The 4026, shown in Fig. 4.22, operates with a logical 0 (GND) level applied to RESET and CK ENABLE, and a logical 1 level applied to DISPLAY ENABLE. The counter advances by one at each positive-going edge

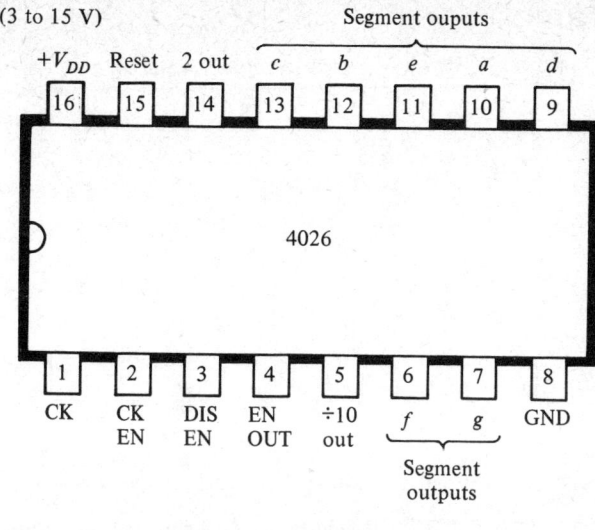

Fig. 4.22. CMOS 4026 decade counter with 7-segment decoded output.

of the CK pulse. The segment outputs are at a logical 1 level when that segment is to be illuminated, the output drive current being limited to about 1.2 mA at 5 V.

The counter may be reset to zero by applying a logical 1 level, and must be at logical 0 for the count to continue. To conserve display energy, the display may be turned off by applying a logical 0 level to the DISPLAY ENABLE terminal.

(c) *Dual-synchronous* counters are most useful for timing applications. Typical devices in this class include the 4518-dual divide-by-10 counter and the 4520-dual divide-by-16 counter.

(d) *Up-down counters* are often referred to as *add/subtract* counters, since they are able to add to, as well as subtract from, the stored count. This type of counter is widely used in frequency synthesizers. Typical devices include the 4510 (divide-by-10 BCD up/down counter) and the 4526 (divide-by-10 or divide-by-16 counter).

(e) *Variable modulus counters* include the 4018 (divide-by-2 to 10), the 4522 (decimal divide-by-N), and the 4526 (binary divide-by-N).

The 4018 is clocked on the positive edge of the CK pulse, and operates at frequencies of 2.5 MHz at 5 V and 5 MHz at 10 V. The pin-outs of the 4018 are shown in Fig. 4.23. A selection of connection diagrams for different division applications using the 4018 are shown in Fig. 4.24.

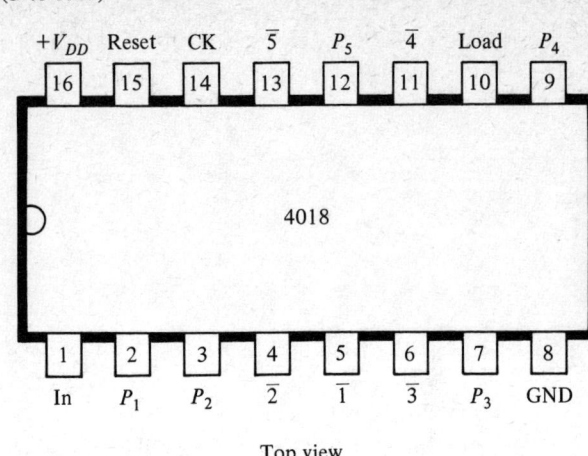

(3 to 15 V)
+V_{DD} Reset CK $\overline{5}$ P_5 $\overline{4}$ Load P_4
16 15 14 13 12 11 10 9
4018
1 2 3 4 5 6 7 8
In P_1 P_2 $\overline{2}$ $\overline{1}$ $\overline{3}$ P_3 GND

Top view

Fig. 4.23. CMOS 4018 – divide-by-2 to 10 synchronous counter.

+V_{DD}
f_{in}
16 15 14 13 12 11 10 9
4018
1 2 3 4 5 6 7 8
$f_{in}/2$
$f_{in}/4$
$f_{in}/7$
$f_{in}/10$

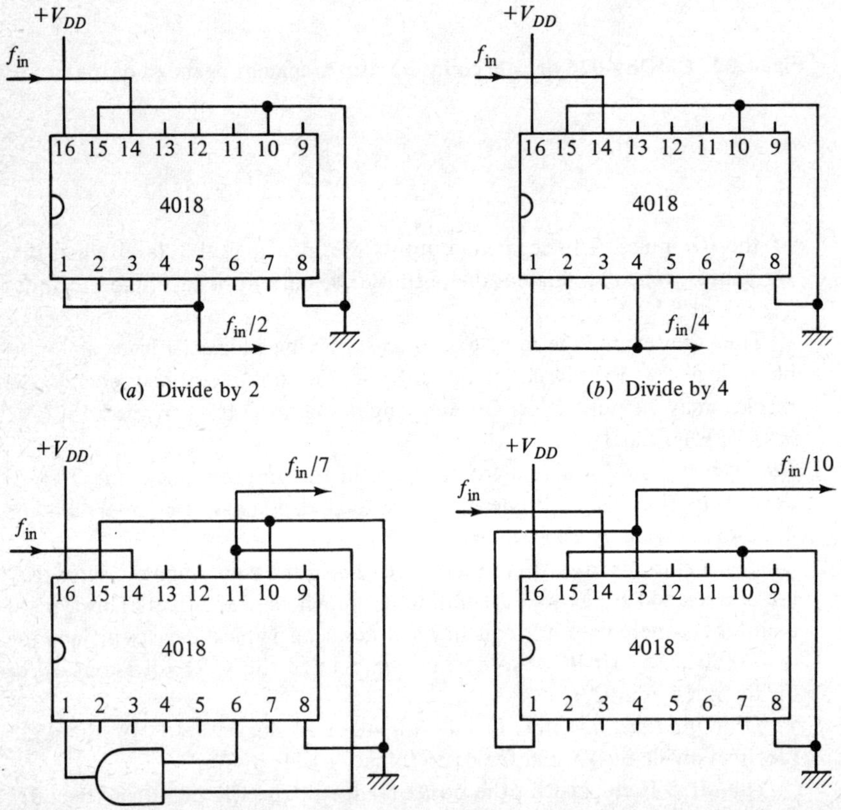

(*a*) Divide by 2

(*b*) Divide by 4

(*c*) Divide by 7

(*d*) Divide by 10

Fig. 4.24. Applications of the 4018.

EXERCISES TO CHAPTER 4

1. Briefly explain the differences between synchronous and asynchronous counters.

2. Explain why the asynchronous 'ripple-through' counter is slower than the synchronous counter.

3. Explain the meaning of the term 'propagation delay' and, with the aid of diagrams, illustrate its effect in an asynchronous counter.

4. Sketch the logic diagram of a bistable element connected to operate as a divide-by-two arrangement. Identify whether the network is clocked by positive or negative edges, and sketch the input and output waveforms.

5. Sketch the logic diagram of a 3-bit asynchronous binary counter and explain its operation.

6. Sketch the logic diagram of a 4-bit synchronous binary counter.

7. Sketch the time-related waveforms for a 4-bit asynchronous counter and hence draw the truth table.

8. Explain how a 4-bit asynchronous binary counter may be converted into a down-counter.

9. Derive the truth table for a decade counter.

10. Sketch the logic diagram of a decade counter, and explain its operation.

11. Describe, with the aid of sketches, how the count in any counter can be curtailed to any desired value.

12. Sketch the logic diagram of a 4-stage binary counter which has its count curtailed to a scale-of-seven counter.

13. Explain the need for a decade counter.

14. Explain how a 4-bit binary counter can be used as a divide-by-16 frequency divider.

15. Sketch the logic network of an 8421 BCD counter, label all the connections, and sketch the time-related waveforms of all the stages throughout a full count sequence.

16. Sketch the logic diagram of a scale-of-six counter using the SN 7490 decade counter.

17. Describe, with the aid of sketches, how an asymmetrical waveform may be converted to a symmetrical 1 : 1 mark/space ratio waveform.

18. The CMOS 4045 is listed as a 21-stage counter. Determine the highest order of frequency division that may be attained.

19. Explain why some CMOS counters have limitations on the availability of their outputs.

20. The clock input to a CMOS 4020 14-stage counter is 1 MHz derived from a crystal-controlled source. Determine the *lowest* frequency signal output which may be obtained.

5 Registers

5.1 Introduction

A *register* is made up from a group of bistable elements connected together in such a way as to be capable of storing binary data or counting sequences as logic levels.

A *shift register* is a register which is designed to allow the data to be shifted along the register, either to the right or to the left.

A *ring counter* is a shift register which is connected in the form of a continuous ring.

5.2 Storage registers

Storage registers can be made up by using virtually any of the types of bistable elements discussed in Chapter 3, but those most commonly used in registers are controlled by clock pulses.

A simple 4-bit storage register using *D-type* bistable elements is shown in Fig. 5.1, which is a *parallel-in* and *parallel-out* (PIPO) register.

The data to be stored in this register is applied to the D inputs of the bistable elements as a binary number (all bits simultaneously, i.e., *parallel data in*), and on application of a clock pulse, the data is stored in the register and is readily available at the Q outputs of the bistable elements as a parallel-out binary

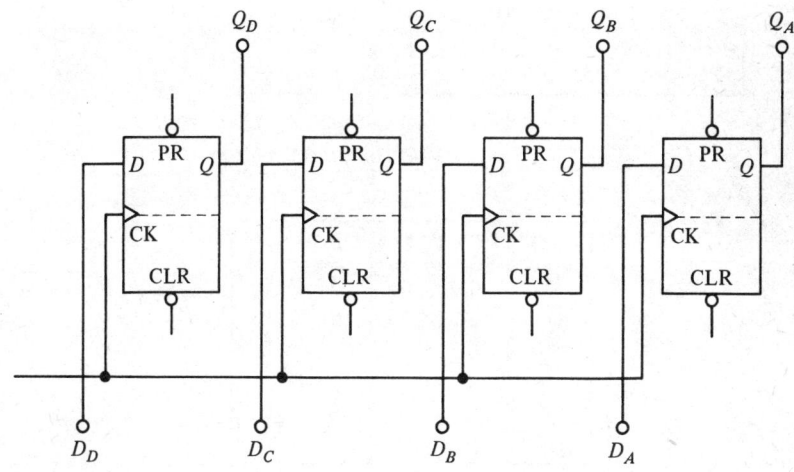

Fig. 5.1. Simple 4-bit storage register.

number. Bistable elements having preset and clear facilities, such as the SN 7474 dual *D*-type bistable, may use these connections to enter the data into the register.

Note: If NO input is applied to the *D* input in the SN 7474, then a logical 1 level is normally assumed.

5.3 Shift registers

The simple storage register can be modified to a shift register by connecting the output of one bistable element to the input of the next, and so on, through the register. (A simple 4-bit shift-right register using *D*-type bistable elements is shown in Fig. 5.2.) Therefore, if it is required that a logical 1 level is entered at the serial input end, and this logical 1 level is to be shifted along the register one bistable element at a time, i.e., at each positive-going edge of the CK input pulse, then it is necessary initially to apply a logical 1 level to the *D* input of the first bistable at a time which is coincident with the active edge of the first clock pulse. At the end of the first clock pulse, the *Q* output of the first bistable becomes logical 1, so that the *D* input of the first stage must now be set to logical 0 for the remaining sequence of operations. Each successive clock pulse causes the logical 1 signal to be propagated *one bistable element to the right* as shown in the truth table in Fig. 5.3.

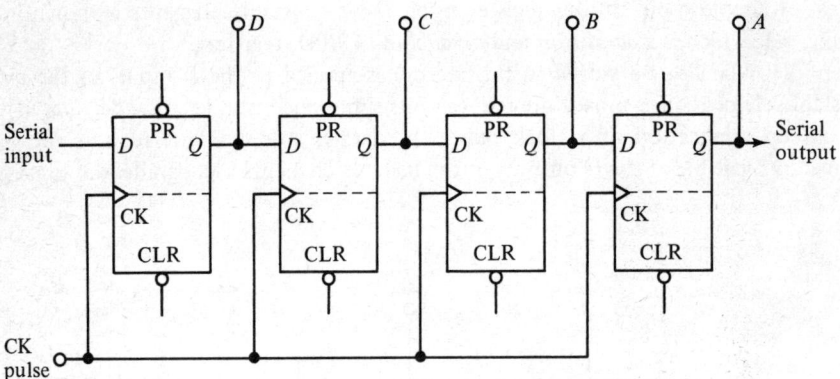

Fig. 5.2. Simple 4-bit shift-right register.

Clock	D	C	B	A
0	0	0	0	0
1	1	0	0	0
2	0	1	0	0
3	0	0	1	0
4	0	0	0	1
5	0	0	0	0

Fig. 5.3. Truth table for shift-right register.

The data may be read as parallel-out data from the Q outputs of each of the bistable elements at any instant, or can be read as *serial-output* data from the Q output of the final stage.

Data may be entered in parallel in the register shown in Fig. 5.2 by applying it as a complemented parallel binary number to the preset inputs of the bistable elements.

5.4 Data control in shift registers

We have seen from the simple register and the shift register that data can be fed either in parallel (all digits applied simultaneously as a binary number) or in serial form (digits sequentially applied to the input of the first bistable element). Similarly, data can be read out from the register in parallel (all digits read simultaneously) or in serial form (digits read out sequentially from the output of the last bistable element).

If the bistable elements used in the register do not have preset and clear facilities, then parallel data input may be controlled by the use of additional logic circuitry, as shown in Fig. 5.4.

When the ENABLE parallel data in is at a logical 1 level, the inputs to the D-type bistable elements are the same as the required data to be stored (D, C, B, A), so that on application of a clock pulse, that data is stored in the register. If the ENABLE parallel data in is at logical 0, then the input to each bistable element is the same as the Q output of the previous one.

If the bistable elements used in the register are provided with preset and clear facilities, then a register may be constructed with parallel data in, parallel-out, or any combination of serial/parallel data, SISO, SIPO, PISO, and PIPO, as shown in Fig. 5.5.

Once the data has been fed in it will be shifted one place to the right at each clock pulse, i.e., *shift-right* register.

5.5 Reversible shift register

The registers considered so far are only capable of shifting data to the right. The data may be shifted either to the right *or* to the left by the use of a logic control network, as shown in Fig. 5.6, in which the 4-bit reversible shift register has facilities for parallel input/output (PIPO) and serial input/output (SISO). The register shown in Fig. 5.6 may be constructed using $2 \times$ SN 7474 dual D-type bistable elements, $2 \times$ SN 7400 quad 2-i/p (input) NAND gates, $2 \times$ SN 7450 dual 2-wide 2-i/p AND-OR-INVERT gates, and a $1\frac{1}{6} \times$ SN 7404 hex inverter.

The 4-bit binary number is applied to the *parallel data in* terminals and loaded into the register by applying a logical 1 level to the *data load* terminal. Logical signals may now be applied to the *shift control*, and the contents of the register will then be shifted (to the right or left) on application of the CK pulses. Alternatively, the 4-bit number may be fed in serially to either the *serial input-shift left* terminal or to the *serial input-shift right*. Again, the data will be shifted through the register as clock pulses are applied to CK.

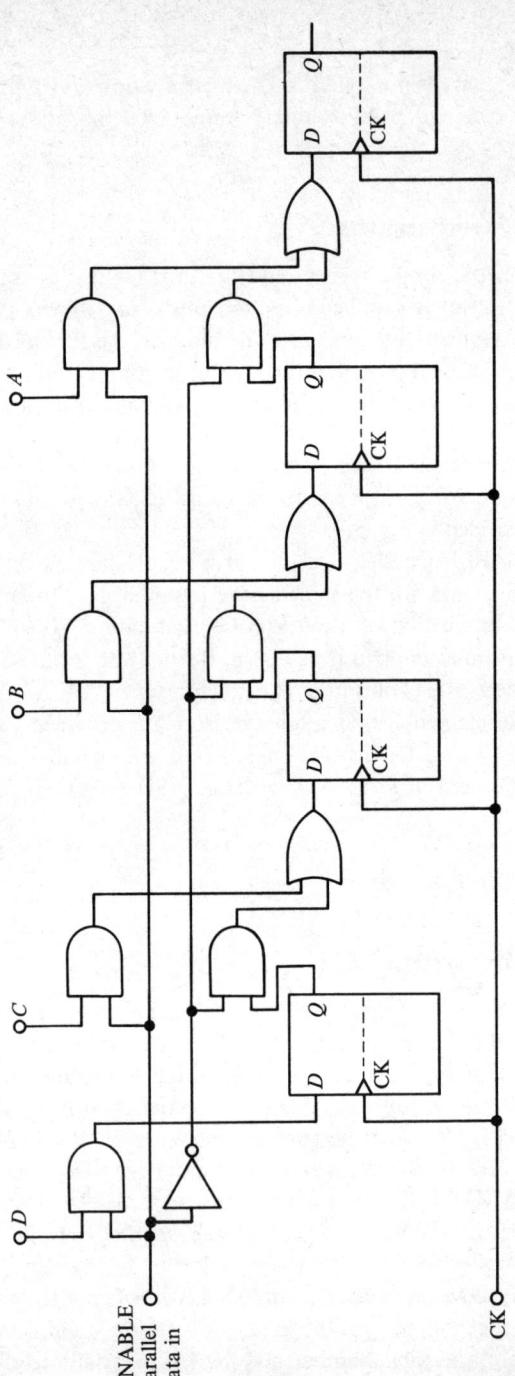

Fig. 5.4. Parallel data input control in register.

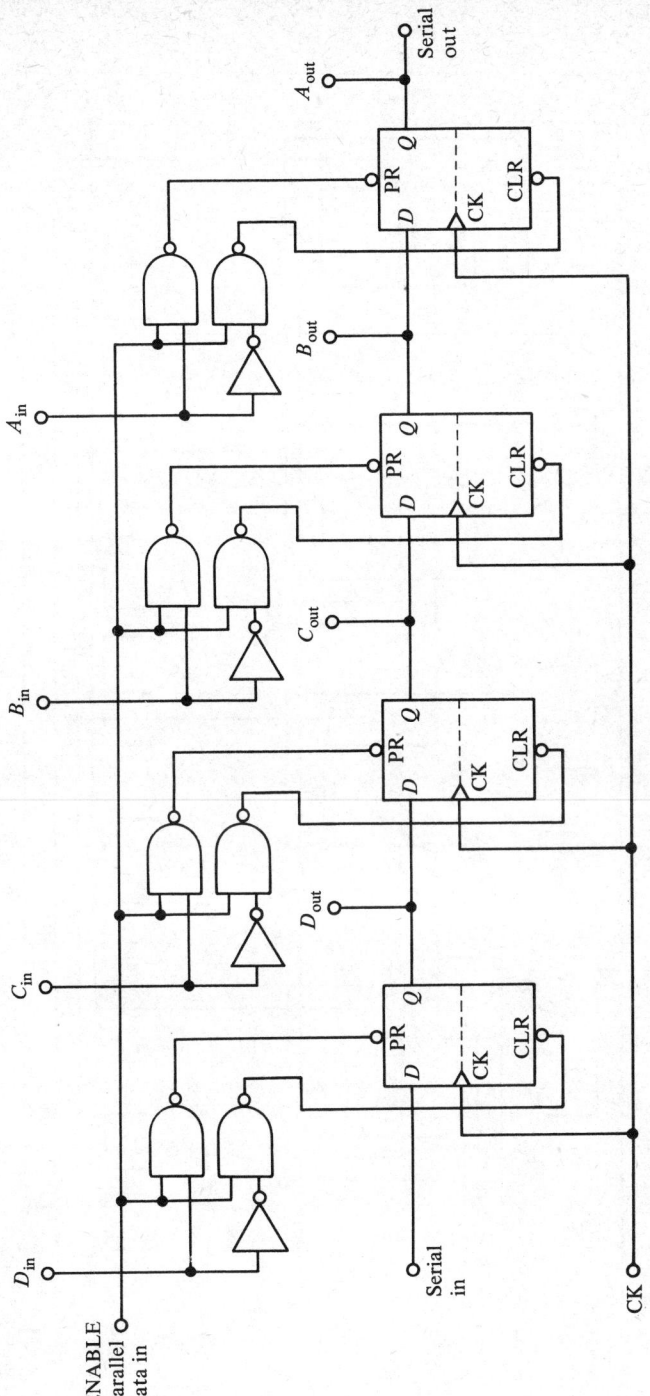

Fig. 5.5. Serial/parallel shift register with parallel data control.

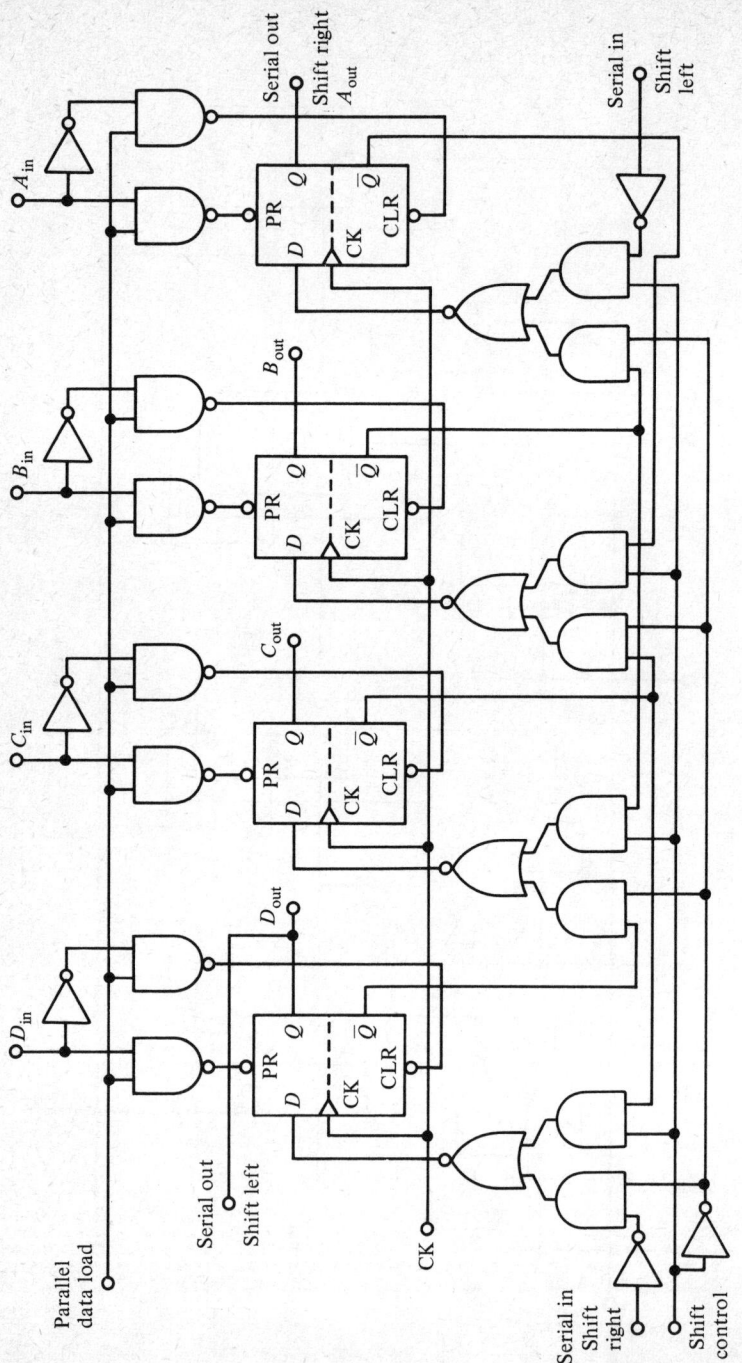

Fig. 5.6. 4-bit reversible shift register.

5.6 TTL shift registers

Many registers are available in the TTL family of integrated circuits. Examples will be considered of an 8-bit SISO shift-register and a 4-bit PIPO/SIPO shift register.

(a) *SN 7491A–8-bit shift register.* The pin-out diagram is shown in Fig. 5.7. Single-rail data and input control are gated through inputs A and B, and an internal inverter to form the complementary inputs to the first bit of the shift register. The clock pulse inverter/driver causes the data to be shifted on the positive-going edge of CK. This device can be operated at clock frequencies up to about 18 MHz.

   The function table is shown in Fig. 5.8.

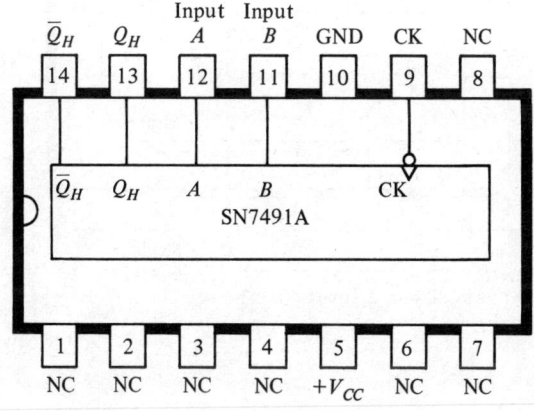

NC = not connected

Top view

Fig. 5.7. SN 7491A 8-bit shift register.

Inputs at t_n		Outputs at t_{n+8}	
A	B	Q_H	$\bar{Q}_H$
H	H	H	L
L	X	L	H
X	L	L	H

H = high logic level

L = low logic level

X = irrelevant logic level

t_n = reference time, with CK = low

t_{n+8} = time after 8 CK pulses (low to high)

Fig. 5.8. Function table for SN 7491A.

(b) *SN 7495A—4-bit parallel-access shift register.* The pin-out diagram is shown in Fig. 5.9. This is one of a group of 4-bit registers having parallel and serial inputs, parallel outputs, mode control, and two clock inputs. *Parallel loading* is effected by applying the 4-bit binary number and applying a logical 1 (high) level to the mode control. Data is loaded into the register and appears at the outputs after the negative-going edge of CK_2. Serial data input is inhibited during parallel loading.

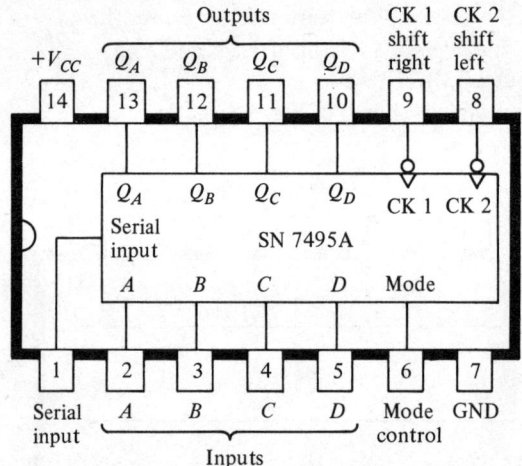

Fig. 5.9. SN 7495A 4-bit parallel-access shift register.

Shift-right is achieved on the positive-going edge of CK_1 while the mode control is at a logical 0 (low) level. Shift-left is achieved on the negative-going edge of CK_2 while a logical 1 is applied to the mode control. The SN 7495A can be operated at frequencies up to about 36 MHz.

The SN 7495A shift register may be used in an application of converting a 4-bit parallel binary number to a serial number, as shown in Fig. 5.10. The 4-bit binary number is applied to the parallel input, and the initial input is set to logical 0 to load the register. The data is shifted out one bit at a time as CK pulses are applied.

5.7   The ring counter

The ring counter is basically a shift register whose input is obtained from its output, as shown using D-type bistable elements in Fig. 5.11.

If all the Q output states are initially set at logical 0, and then a logical 1 is loaded into the first bistable element, the first positive edge of the clock pulse will cause all the Q output states to shift one place to the right. Thus, the logical 1 in bistable D appears in bistable C, and the logical 0 in bistable A is fed back to

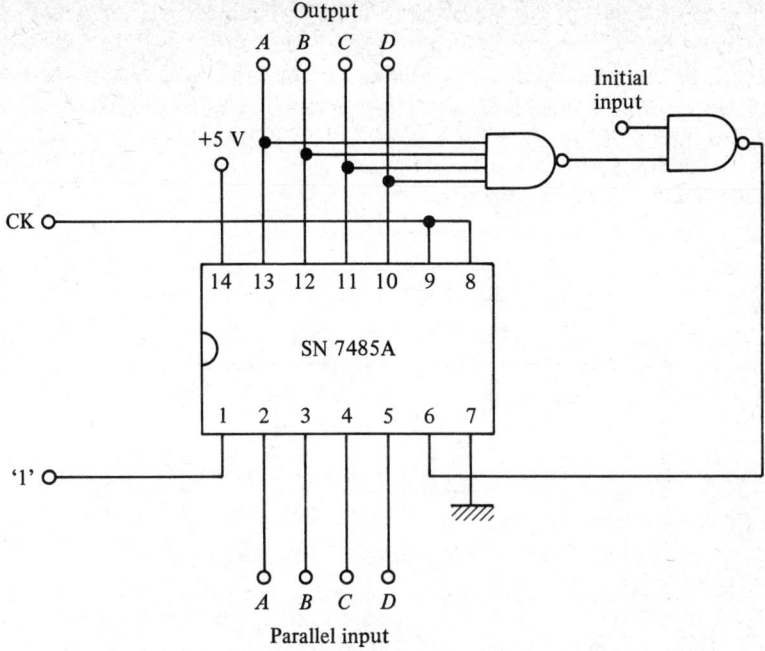

Fig. 5.10. SN 7495A parallel (4-bit)/serial converter.

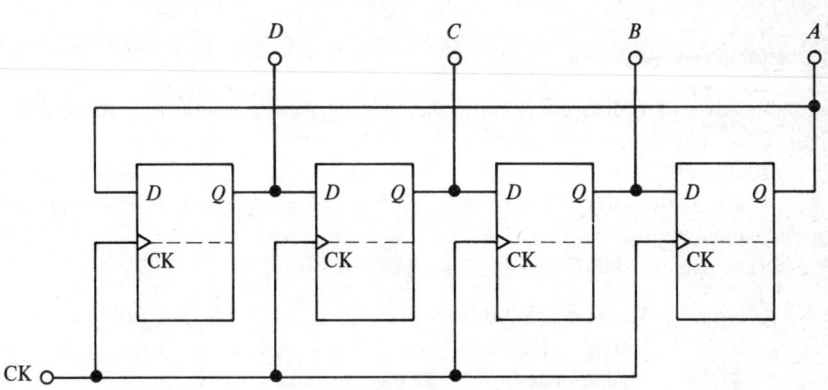

Fig. 5.11. The simple ring counter.

bistable D. Therefore, the logical 1 circulates around the register, one bistable at a time, as the clock pulses are applied.

The ring counter is relatively easy to decode, e.g., in the simple ring counter shown, when Q_D is logical 1 the count is zero, when Q_C is logical 1 the count is one, when Q_B is logical 1 the count is two, etc. It can therefore be seen that the cycle length of the code generated is four for the simple ring counter shown, and ten bistable elements would be required for a decade counter. The basic ring counter is therefore not very economical in the use of bistable elements.

The cycle length of the ring counter may be doubled by feeding back the output from the $\bar{Q}_A$ output (instead of Q_A), which is then called a *twisted-ring* counter. However, the twisted ring counter is more difficult to decode, as shown in the truth table in Fig. 5.12, assuming that the Q output states of all the bistable elements are initially at logical 0.

Denary	Q_D	Q_C	Q_B	Q_A	
0	0	0	0	0	
1	1	0	0	0	
2	1	1	0	0	
3	1	1	1	0	Repeat
4	1	1	1	1	
5	0	1	1	1	
6	0	0	1	1	
7	0	0	0	1	

Fig. 5.12. Truth table for the twisted ring counter.

To ensure that the bistable elements are initially set to the required conditions, a logical network may be used. The twisted-ring counter may be constructed using *D*-type bistable elements, as shown in Fig. 5.13, or using *J–K* bistable elements as shown in Fig. 5.14.

5.8 CMOS registers

As in the case of CMOS counters, CMOS registers are available with more stages than TTL. The speed of operation of CMOS devices is generally slower than TTL devices, and is also dependent upon the supply voltage. It is *always advisable* to consult manufacturer's data to ascertain the operating limitations of a particular device *before* use.

A selection of CMOS registers is listed as follows:

4006	18-stage SISO shift register
4014	8-stage PISO shift register
4015	dual 4-stage SIPO shift register
4021	8-stage PISO shift register
4031	64-stage SISO shift register
4035	4-stage PIPO shift register
4042	quad latch storage register
4508	dual 4-stage storage register

The pin-out connections of the CMOS 4014 8-stage shift register are shown in Fig. 5.15, which is a device which may be used as a 6-, 7-, or 8-stage shift-right register, either as SISO or as PISO.

When operated in the SISO mode, the load input is connected to logical 0 (GND). Data applied to the IN terminal will be shifted into the first stage on the

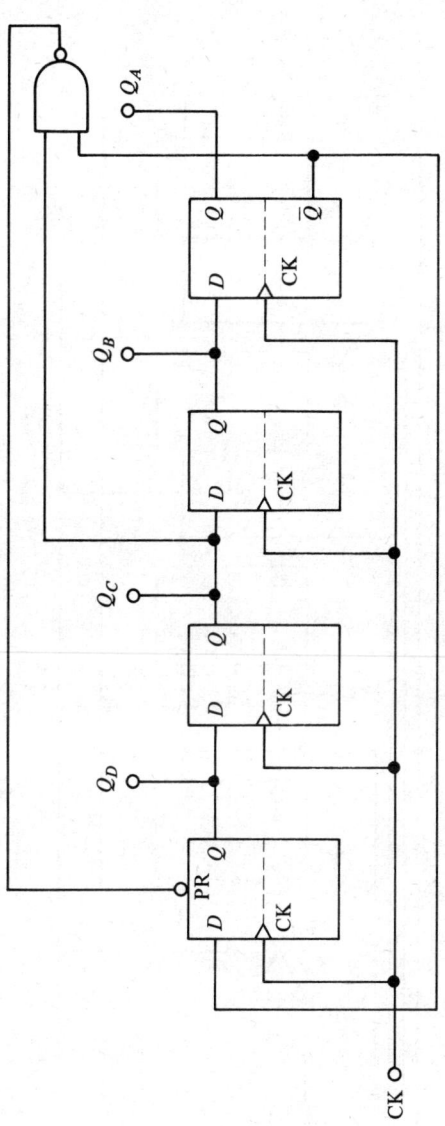

Fig. 5.13. Twisted ring counter using *D*-type bistable elements.

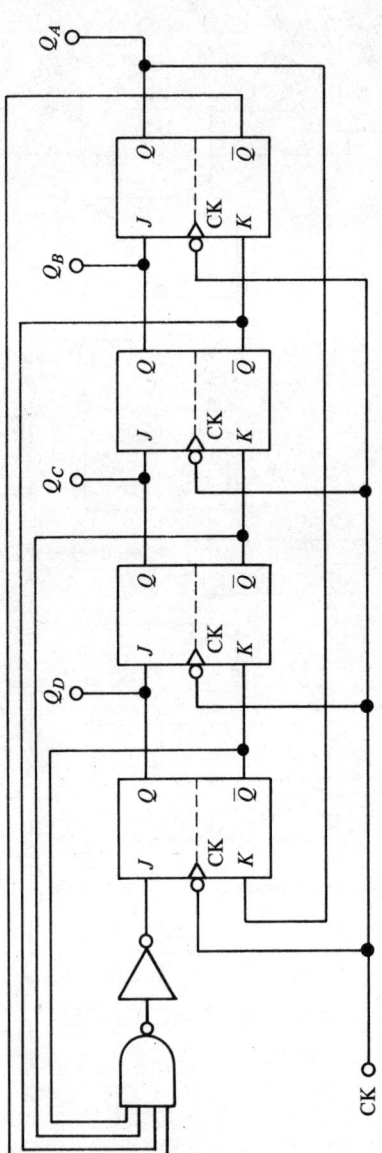

Fig. 5.14. Twisted ring counter using J–K bistable elements.

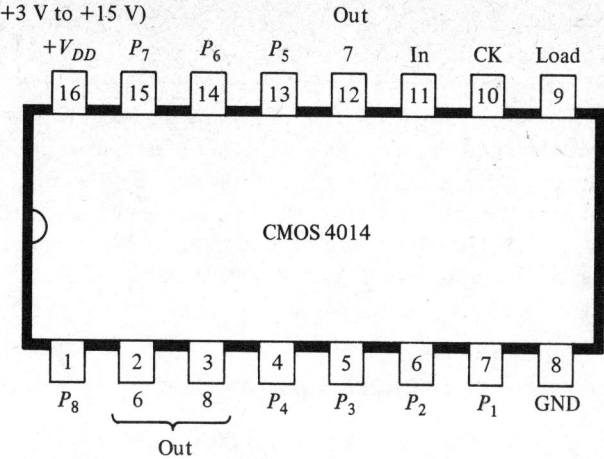

Fig. 5.15. CMOS 4014 8-stage shift register.

first positive edge of the clock pulse CK. After eight successive clock pulses, the data originally applied to IN will appear at output 8 (pin 3). Further clock pulses will lose this data unless it is recirculated or cascaded to another stage.

To operate the device in the PISO mode, the 8-bit binary number is presented in parallel to the P_1–P_8 terminals, and a logical 1 level applied to the LOAD input coincident with a positive edge of the clock to synchronously load the data. The LOAD input is then set to logical 0 for normal register operation.

The maximum clock frequency is about 5 MHz at 10 V and 2.5 MHz at 5 V.

The pin-out connections of the CMOS 4031 64-stage shift register are shown in Fig. 5.16, which is a fully static shift register arranged as 64 SISO stages.

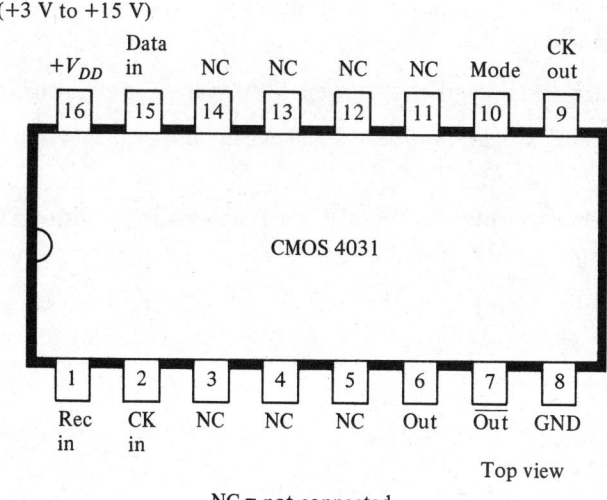

NC = not connected

Fig. 5.16. CMOS 4031 64 stage SISO shift register.

Normally, data is applied to DATA IN, and the MODE input is set to logical 0 (GND). The data enters the register on the positive edge of the clock. After 64 clock pulses, the data appears at the OUT terminal (pin 6), and its complement $\overline{\text{OUT}}$ (pin 7). These outputs are each capable of driving one TTL load.

If the MODE input is set to logical 1, data presented to the REC IN (re-circulating) input is entered to the register on the next positive edge of the clock. For recirculation, the REC IN terminal may be connected to OUT (pin 6). Alternatively, the MODE control can be used to select the serial data at REC IN (MODE = logical 1) or applied to DATA IN (MODE = logical 0).

The maximum clock frequency is about 4 MHz at 10 V and 2 MHz at 5 V.

EXERCISES TO CHAPTER 5

1. Sketch the logic diagram of a simple 4-bit register, and briefly explain its operation.

2. Explain the meaning of the terms *serial input* and *parallel input* with reference to a register.

3. Sketch the logic diagram of a shift register in which data may be entered in parallel and read out as a parallel number.

4. Explain the meaning of the abbreviations SISO, SIPO, PISO, and PIPO.

5. Describe, with the aid of sketches, how a simple shift register may be converted into a reversible shift register.

6. Explain the meaning of the terms *left* and *right shift capability*, and *shift mode control*, as applied to registers.

7. Explain how a register may be used to store digital data in the form of binary numbers, and describe how the register may be used to manipulate that data.

8. Sketch the logic diagram of a 4-bit SIPO using *D*-type bistable elements.

9. State typical propagation delays and maximum frequency of clock pulses for TTL and CMOS registers.

10. Explain the advantages of CMOS registers compared with TTL registers.

6 Logic circuit families

6.1 Introduction

Electronic logic elements have evolved through a number of stages, beginning with systems consisting of *diode* AND and OR gates. Advances in semiconductor technology fostered rapid developments in electronic logic circuitry of the active type, and various circuits were produced. The first *integrated* logic elements were simply translations of discrete component circuits directly into silicon circuits. The earliest types were, in fact, composed of several silicon chips with wire interconnections. As integrated circuit techniques developed, the design approach changed and the circuits began to be designed to suit the manufacturing technology, instead of being duplicates of discrete component prototypes. Once it was realized that circuit complexity was not a limiting factor, the way was open for the production of high-performance, complex circuit elements.

6.2 Choice of logic family

The choice of a logic family for a particular application is generally determined by consideration of the following factors:

1. Speed of operation
2. Noise immunity
3. Power dissipation
4. Operating temperature range
5. Type of package
6. Cost
7. Availability

These factors are not necessarily given in any particular order of priority, the individual application generally dictates in which order the various factors must be considered.

6.3 Speed of operation—propagation delay

The speed of a logic gate is defined by its *propagation delay*, i.e., the time taken for a logical signal to pass through the gate from input to output. One of the contributing factors for this propagation delay may be illustrated by considering a diode, in which the applied voltage is changed from forward bias to reverse bias. The current in the forward direction does not in fact fall immediately to zero (or, at least, to the leakage current value), since the charge carriers must first recombine with their parent atoms, and thus disappear. This causes a pulse of *reverse* current to flow (as shown in Fig. 6.1), and which takes time to decay to

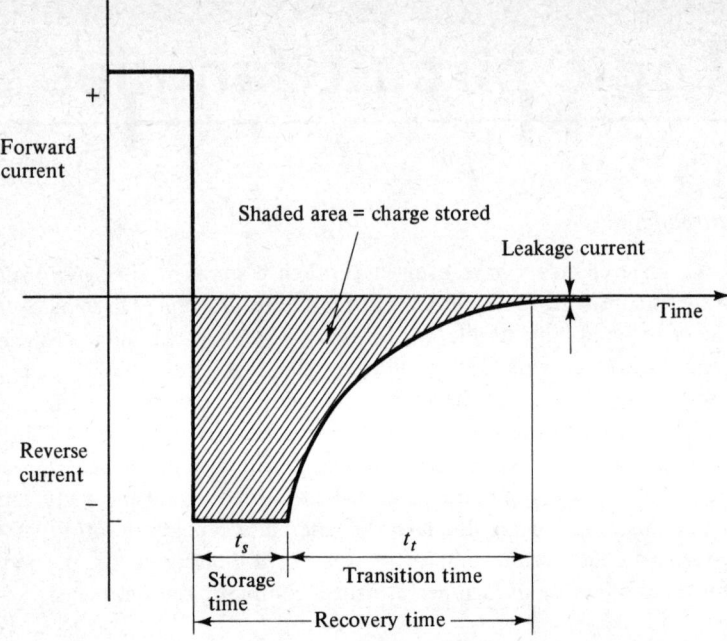

Fig. 6.1. Charge storage in the diode.

the leakage value. During this time, the reverse current flow causes an electric charge to be stored in the diode junction. Once the charge carriers have been swept away from the junction (*storage time, t_s*), the reverse current decays to the leakage value (*transition time, t_t*). The total delay (*recovery time*) represents the propagation delay of a signal switched by a diode.

Efforts to improve the switching performance of bipolar devices led to the use of the Schottky diode, in which a barrier is formed between a metal and *n*-type semiconductor. The current flow in these devices does not depend in any way on minority charge carriers so that the effects of charge storage are almost eliminated. Schottky junctions are used in diodes and transistors where propagation delays of the order of a few nanoseconds are required. The circuit symbols most widely used throughout industry for the Schottky diode and transistor are shown in Fig. 6.2.

Some digital equipment, e.g. machine tool control systems, operates at a relatively slow rate where a propagation delay of 1 ms may be acceptable. However, modern high-speed digitial computers and micro-computers may require propagation delays approaching 1 ns. Most integrated circuit (IC) logic families manufactured today have propagation delays between 2 and 100 ns.

The asymmetrical switching characteristics of ICs is such that a *low-to-high* logical transition signal at the gate input has a different delay from that of a *high-to-low* logical transition at the gate input. These logic levels are referred to as the *threshold levels* and specifications generally include a *minimum high* threshold level and a *maximum low* threshold level for a particular logic family.

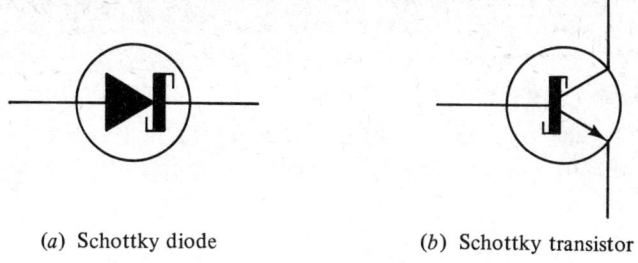

(*a*) Schottky diode   (*b*) Schottky transistor

Fig. 6.2. Symbols used for Schottky devices.

Typical propagation delays are usually taken as the average of the two delays stated above, as shown in Fig. 6.3 for an inverter (NOT gate), and specified at the 50 per cent signal levels.

Stray capacitance at the gate output has a considerable effect on the propagation delay, so that quoted figures generally refer to a given value of capacitance, for example, 15 to 30 pF.

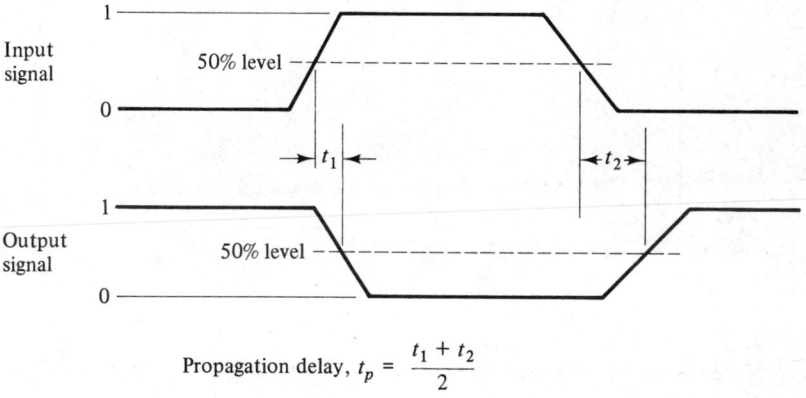

$$\text{Propagation delay, } t_p = \frac{t_1 + t_2}{2}$$

Fig. 6.3. Propagation delay in an inverter (NOT gate).

6.4 Noise immunity

Spurious voltages occurring on signal interconnection paths are termed *noise*. This type of signal can cause erroneous switching of logic gates. Within the system, noise is usually self-generated as a result of 'cross-talk' between signal paths. Logic gates are generally designed to have a built-in immunity to this type of noise, the *noise margins* being defined in terms of the threshold levels. The *low noise margin* is the difference between the maximum low output voltage level and the minimum low input threshold level. The *high noise margin* is the difference between the minimum high output voltage level and the maximum high input threshold level. The smaller of the two values is most generally quoted. Thus:

Noise immunity is the degree to which a logic gate can withstand variations in input levels without causing a significant change in output logical state, i.e., the *d.c. noise margin is the difference between the output voltage and the input threshold.*

Consider the transfer characteristic of a TTL NAND gate (operated as an inverter) as shown in Fig. 6.4. The shaded areas represent *forbidden* values by

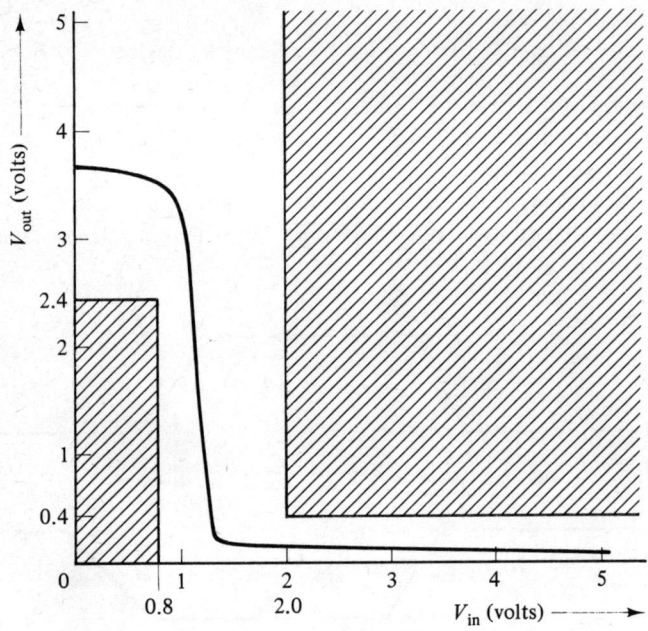

Fig. 6.4. Transfer characteristic of a TTL NAND gate (connected as an inverter).

the specification, which states that *the output of a gate is guaranteed to be less than* 0.4 V *in the logical 0 state, and guaranteed to be greater than* 2.4 V *in the logical 1 state. The input threshold is guaranteed to be between* 0.8 *and* 2.0 V.

Therefore, the *worst-case* noise margins for the TTL NAND gate are:

Noise margin in the LO '0' state = 0.8 − 0.4 = 0.4 V
Noise margin in the HI '1' state = 2.4 − 2.0 = 0.4 V
∴ the guaranteed noise margin   = 0.4 V = 400 mV, in this case

6.5 Fan-in and fan-out

The *fan-in* of a logic gate is the maximum number of separate inputs which may be applied to the gate; this is largely determined by the propagation delay of the gate.

The *fan-out* of a logic gate is the maximum number of basic gate inputs that the gate may supply simultaneously without causing the output logical level to fall outside its specification.

6.6 Power dissipation

It is necessary to know the power dissipation in order to determine the power supply requirements. Most logic gates draw a different current from the supply depending on whether the output is at a logical 1 or a logical 0. Typical values quoted will be the average of the two.

The faster circuits tend to dissipate more power, since they are generally designed with lower values of resistors; also, at faster switching speeds, the charging action of the stray capacitance tends to draw more current. Most logical IC families operate with supply voltages in the region of 5 V, and typical power dissipations range from 1 to 100 mW per gate.

6.7 Operating temperature range

The operating temperature is the ambient temperature in which the device will operate satisfactorily and meet its specification. Two standard ranges are widely used: the *military* (−55 to +125 °C), and the *commercial* (0 to +70 °C). Certain families are available in restricted temperature ranges only.

6.8 Logic families

Unfortunately, integrated circuit manufacturers tend to introduce their own designs of logic circuits with little standardization in mind. Eventually, however, certain types emerge as being more popular and these are duplicated by other manufacturers, thus providing multiple source availability.

It is not possible to draw the circuit diagram of an integrated circuit—even if we should want to do so. However, in order that we may use conventional theories to confirm functions for ourselves, the manufacturers include 'functional' or 'schematic' diagrams *not circuit diagrams*. Unfortunately, we use these diagrams so much that we begin to believe that they are the *actual circuits*.

It has been common practice to classify IC logic families by the circuit configuration of the basic gate technology, the earlier types of which were: *R*esistor *T*ransistor *L*ogic (RTL) and *D*iode *T*ransistor *L*ogic (DTL). These early types have largely been superseded by the following IC logic families:

*T*ransistor-*T*ransistor *L*ogic (TTL);
*E*mitter-*C*oupled *L*ogic (ECL);
CMOS Logic (*C*omplementary *M*etal-*O*xide *S*ilicon).

6.9 Transistor-transistor logic (TTL)

This may be considered as a development of DTL in which the input diodes are replaced by the emitter-base junctions of a multi-emitter transistor as shown in the *functional* diagram in Fig. 6.5.

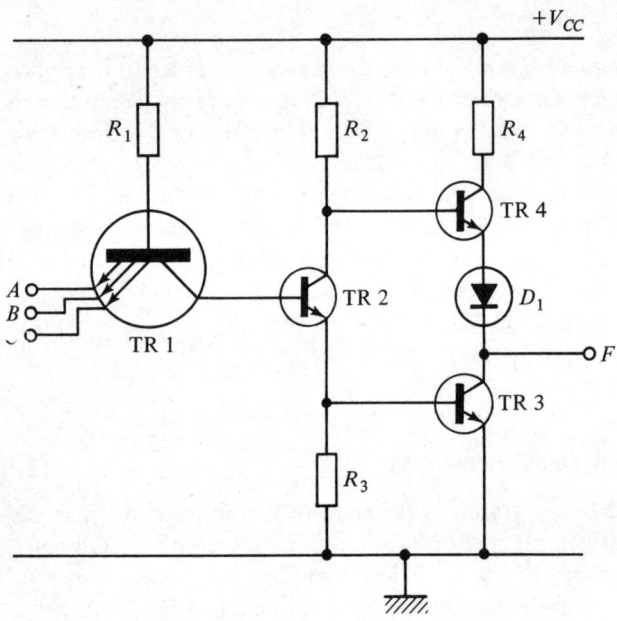

Fig. 6.5. Functional diagram of TTL NAND gate.

The multi-emitter transistor may be easily fabricated using integrated circuit techniques. When all the inputs are at a logical 1 level, the emitter-base junctions of TR 1 are reverse biased, and sufficient current flows through R_1 and the base-collector junction of TR 1 to provide base drive to switch TR 2 hard on, so that it holds TR 3 in saturation (and maintains TR 4 at cut-off), and the output at F is at logical 0. The push-pull nature of this output stage is referred to as a *totem pole* arrangement.

When any or all of the input signals are at a logical 0 level, current flows out of the corresponding emitter of TR 1. This removes the base drive to TR 2, causing it to be cut off, and removing the drive to TR 3, causing it to be cut off. Current now flows through R_2 to drive TR 4 into saturation, and the output signal is at logical 1.

TTL has become a particularly popular range, and as a result many variations exist to reduce power dissipation and propagation delay, and to increase noise margins. Typical figures for standard (*normal*) TTL gates are 10 mW, 10 ns, and up to 1 V. Some of the variations include:

1. *HTTL (high-speed TTL)*, typically 10 mW, 6 ns, and 1 V respectively.
2. *LTTL (low-power TTL)*, typically 1 mW, 35 ns, and 1 V respectively.
3. *STTL (Schottky clamped TTL)*, typically 20 mW, 3 ns, and 0.9 V respectively.
4. *LSTTL (low-power Schottky TTL)*, typically 2 mW, 10 ns, and 0.8 V respectively.

6.10 Sinking and sourcing

When one TTL gate drives another, the limiting currents are specified by the manufacturer's data. When the output of the driving gate is low, i.e., logical 0 level, the path for current flow is as shown in Fig. 6.6(*a*), and the driving gate output (TR 3) is said to *sink* the current, i.e., provide a path to earth. The specification states that the sinking current is a maximum of 16 mA; the standard TTL gates have a nominal fan-out of ten (10), therefore the maximum sinking current for each driven gate (output) is about 1.6 mA.

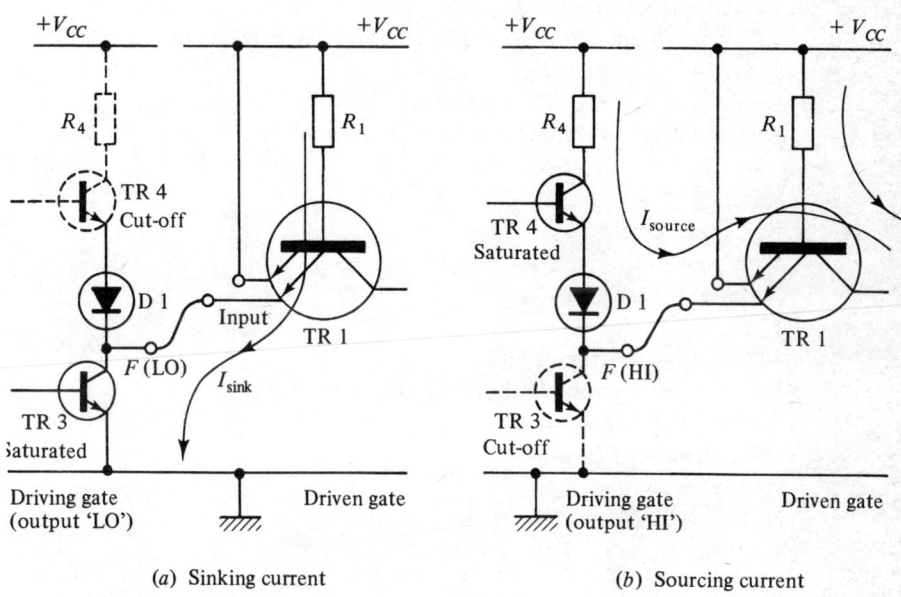

(*a*) Sinking current   (*b*) Sourcing current

Fig. 6.6. TTL sink and source currents.

When the output of the driving gate is high, i.e., logical 1 level, the path of the current flow is as shown in Fig. 6.6(*b*), and the driving gate is said to *source* the current drive for the driven gate (output). The specification states that the source current is a maximum of 400 μA (for nominal fan-out of 10), so that the maximum source current for each output is 40 μA.

Note: The convention used in specifications and data sheets is that all currents are assumed to flow *into* the circuit, i.e., currents flowing into the circuit are labelled as positive, and those which flow out of the circuit are labelled as negative. Therefore, the sinking and sourcing currents and conditions of measurement may be quoted in several different ways depending on whether input or output conditions are being specified:

$$I_{OL} = 16 \text{ mA} \qquad I_{OH} = -400 \, \mu A$$

$$I_{IL} = -1.6 \text{ mA} \quad I_{IH} = 40 \, \mu A$$

6.11 TTL recognition—device numbering

Each part of a device symbolization can be distinctly divided into separate parts, each of which tells us something about the device.

EXAMPLE 6.1

SN 74 H 107 N

SN	Semiconductor network
74	TTL is manufactured to meet two common temperature ranges—

Series 54   −55 to +125 °C—military
Series 74   0 to +70 °C—commercial
Supply voltage tolerances are also different—
Series 54   4.5 to 5.5 V
Series 74   4.75 to 5.25 V

H	High-speed device

Variations of this include—

No letter	Standard TTL
L	Low power
S	Schottky clamped
LS	Low power Schottky

107	Device function (*two* or *three* digits)

107   Dual *J–K flip-flop*

N	Package type

N	14-, 16-, 24-pin DIL Plastic (most common)
J	14-, 16-, 24-pin DIL Ceramic

6.12 Specification—TTL data sheet

A typical data sheet for the basic TTL quad 2-i/p NAND gate is shown in Fig. 6.7. These data sheets are usually prepared to cover *both* the series 54 and the series 74 devices, the main differences between which have already been noted in Sec. 6.11.

The pin-out diagrams are always shown as viewed from the top.

Data sheets always include the supply voltages and temperature ranges, together with the fan-out. This is followed by the *worst-case* parameters, previously discussed in Secs 6.4 and 6.10. Finally, the switching characteristics are given as the propagation delay time *low-to-high* logic level, t_{PLH}, and *high-to-low* logic level, t_{PHL}, with a standard load of 400 Ω resistance and 15 pF capacitance.

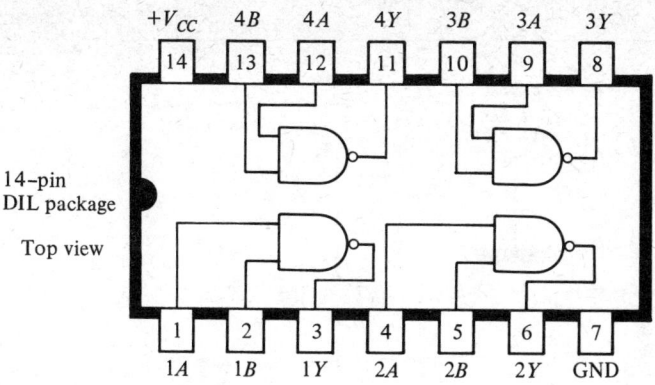

Absolute maximum supply voltage V_{CC} = 7 V
Absolute maximum input voltage V_{in} = 5.5 V
Normal operating supply voltage, 54 series V_{CC} = 5 V ± 0.5 V
Normal operating supply voltage, 74 series V_{CC} = 5 V ± 0.25 V
Operating temperature range, 54 series T_A = −55 to +125°C
Operating temperature range, 74 series T_A = 0 to +70°C
Nominal fan-out, N = 10 $Y = \overline{A \cdot B}$ (positive logic)

Electrical characteristics:

Input voltage: low level V_{IL} = 0.8 V max
                 high level V_{IH} = 2.0 V min
Output voltage: low level V_{OL} = 0.4 V max (typical 0.2 V)
                 high level V_{OH} = 2.4 V min (typical 3.5 V)
Input current: low level I_{IL} = −1.6 mA
                 high level I_{IH} = 40 µA
Output current: low level I_{OL} = 16 mA
                 high level I_{OH} = −400 µA

Switching characteristics:

Load R_L = 400 Ω, C_L = 15 pF
Propagation delay time (low-to-high logic level) t_{PLH} = 22 ns max (typical 11 ns)
Propagation delay time (high-to-low logic level) t_{PHL} = 15 ns max (typical 7 ns)

Fig. 6.7. Data sheet for TTL SN 7400 basic IC.

6.13 Open-collector outputs—TTL

A considerable range of TTL gates and devices are provided with *open-collector outputs*. These are useful in those applications where increased driving capability may be required and/or in those applications where the load is supplied from a higher level of supply voltage, e.g., interface circuits.

The *functional* circuit of an open collector 2-i/p NAND gate is shown in Fig. 6.8, from which it can be seen that the output logic state cannot be determined until an *external pull-up resistor* has been connected between the output terminal and the required supply voltage, which may not be $+V_{CC}$.

Open-collector *buffer* and interface gates can *sink* up to 40 mA, and are available for 15 or 30 V output voltages.

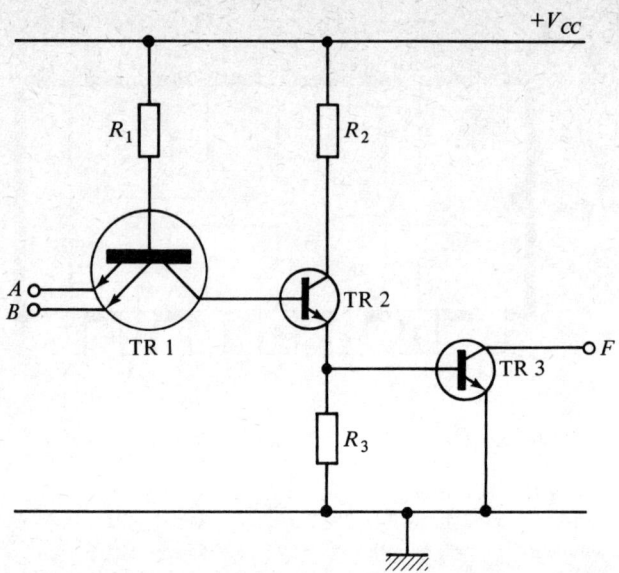

Fig. 6.8. Open-collector TTL NAND gate.

A suitable value of pull-up resistor must be calculated which will ensure that sufficient current flows to the load (and to any paralleled gates) when the output of the gate is high. In addition, the resistor value must be such that sufficient sink current from the TTL load will not cause the output of the gate to rise above the low level of voltage, when in the low logical state.

$$\therefore R_p = V_R/I_R$$

where V_R = voltage drop across pull-up resistor and I_R = current through resistor, including load current and gate current, in each of the logic states.

A typical application of open-collector output devices is the SN 7447A BCD-to-7-segment decoder/driver which features active-low, open-collector, 15 V outputs.

Alternatively, if a device such as a stepper motor runs from a 12 V d.c. supply it requires logic levels which may be as follows:

HI   +7.5 to +12 V
LO   0 to +4.5 V

These signals may be derived by using part of an SN 7407 or SN 7417 hex buffer with open-collector outputs and pull-up resistors as shown in Fig. 6.9.

6.14 Tri-state outputs—TTL

The logical outputs of the gates so far considered are either in a high (HI) state or in a low (LO) state. When it is required for several, or many, digital devices to share a common line, or *bus* system, as in microcomputer systems, then it is

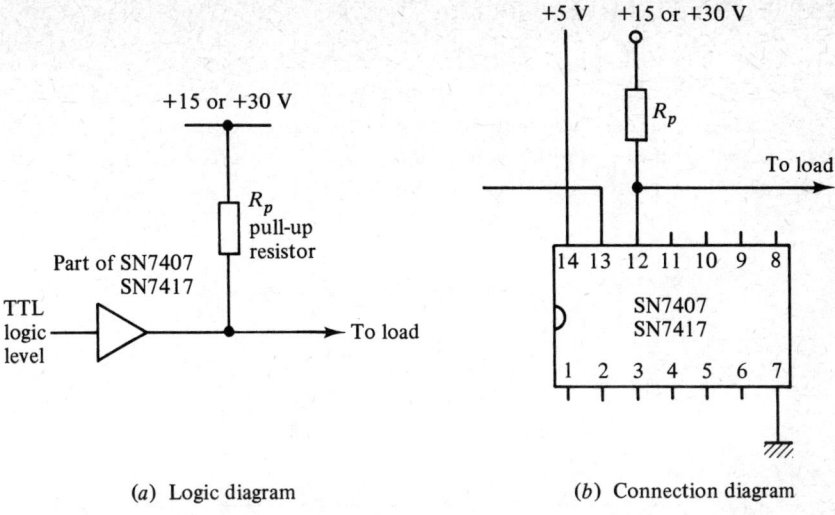

(a) Logic diagram

(b) Connection diagram

Fig. 6.9. Open-collector buffer/driver.

necessary to be able to isolate those devices not currently required. This may be accomplished by using logic devices which have *three* different outputs. In addition to the normal HI and LO outputs, a *high impedance* state can be produced, which provides effective isolation.

Several tri-state (or 3-state) devices are available in the TTL family, including the SN 74125 quad bus buffer and the SN 74126 quad bus buffer, one gate of each being shown in Fig. 6.10. The truth table for each is also shown in Fig. 6.10.

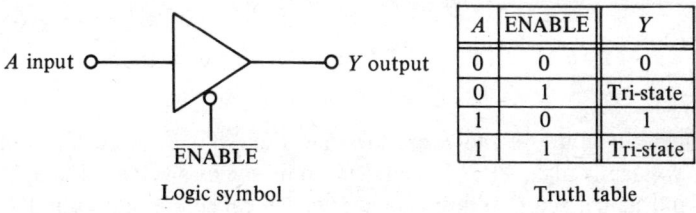

A	ENABLE	Y
0	0	0
0	1	Tri-state
1	0	1
1	1	Tri-state

Logic symbol   Truth table

(a) Bus buffer with tri-state outputs (one of SN 74125)

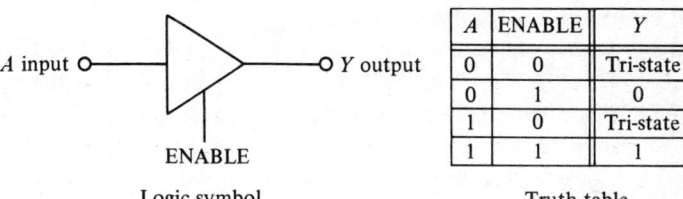

A	ENABLE	Y
0	0	Tri-state
0	1	0
1	0	Tri-state
1	1	1

Logic symbol   Truth table

(b) Bus buffer with tri-state outputs (one of SN 74126)

Fig. 6.10. Tri-state outputs.

6.15   Emitter-coupled logic (ECL)

This logic family differs in principle from most other types in that the transistors are not all operated in saturated conditions. This allows the logic gates to operate more quickly, making them suitable for certain digital computing applications. A *functional* circuit of an ECL gate which is capable of performing both the OR function and the NOR function is shown in Fig. 6.11.

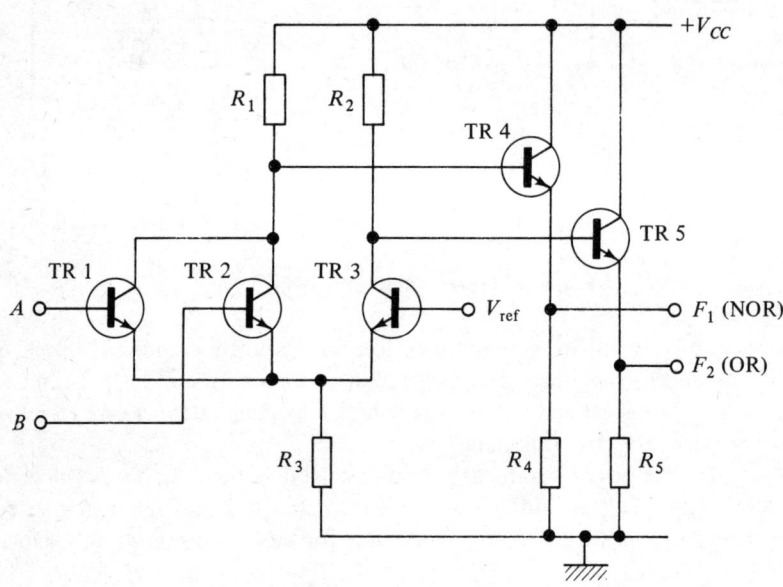

Fig. 6.11. ECL OR/NOR gate functional diagram.

When both inputs A and B are low, TR 1 and TR 2 are cut off and their collector voltage is high. TR 3 is switched on by the bias voltage V_{ref} applied to its base, and its collector voltage is low. The high collector voltages of TR 1 and TR 2 cause TR 4 to switch on, which causes the output at F_1 to go high. At the same time, the low collector voltage of TR 3 is insufficient to switch on TR 5, and the output at F_2 is low.

When any input is high, the emitter-base voltage of TR 3 is reduced, due to the voltage drop across R_3, which causes TR 3 to be cut off, and the collector voltage of TR 3 becomes greater than the collector voltages of TR 1 and TR 2. Under these conditions the output at F_2 is high and the output at F_1 is low. Therefore the logical NOR function is obtained at output F_1, and the logical OR function is obtained at output F_2.

Typical propagation delays of 2 ns are available, with power dissipation of 25 mW and high fan-out of about 30, but noise immunity is low at about 0.2 V. Higher-speed circuits are available at about 1 ns.

6.16 CMOS logic

This logic family is based on *both* n-channel and p-channel MOS devices being used in a complementary symmetrical arrangement. The main advantages of CMOS techniques are (a) small area of fabrication, and (b) low power dissipation. These advantages lend themselves to large-scale integration (LSI) such as some microprocessors and their associated ICs.

The functional circuit of a CMOS inverter (NOT gate) is shown in Fig. 6.12(*a*), and the transfer characteristic for the gate is shown in Fig. 6.12(*b*). When the

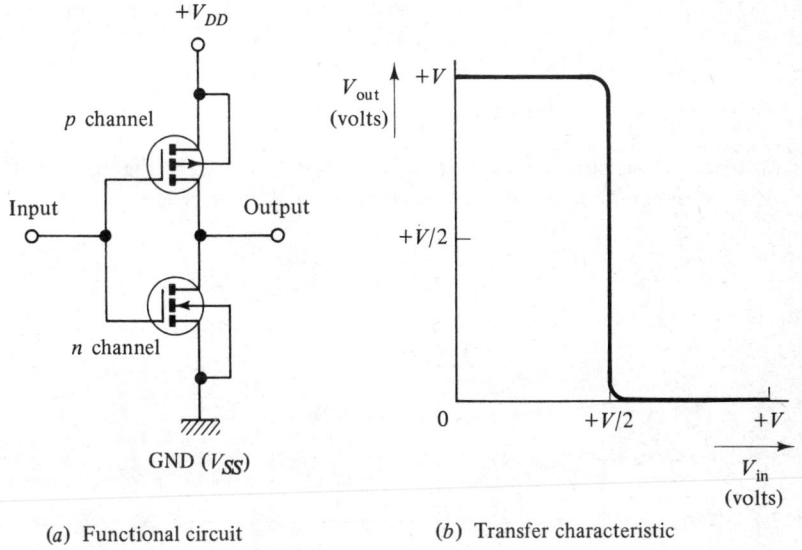

(*a*) Functional circuit   (*b*) Transfer characteristic

Fig. 6.12. CMOS inverter.

input is at a low logical level, the p-channel device is ON and acts as the load resistor for the n-channel device, which is OFF; the output is therefore almost +V (i.e., logical 1). When the input is at a logical 1, the n-channel device is ON, acting as a load resistor for the p-channel device, which is OFF; the output is at a logical 0 level. Note that the output changes state at exactly the point where the input voltage is *half* of the supply voltage.

The transfer characteristic shows that the *actual* values for the high and low states are very nearly $+V_{DD}$ and very nearly 0 V respectively, which means that CMOS circuits have a very high noise immunity, typically 20 per cent of the supply voltage. This family can operate with supply voltages ranging between +3 and +15 V, but for applications such as electronic watch circuits the supply requirements are only about 5 μA at between 1 and 1.5 V, and special care must be taken to reduce the threshold voltage levels. Typical logic levels are: 5 V supply, logical 0 = 0 to +1 V, logical 1 = +4 to +5 V; 10 V supply, logical 0 = 0 to +2 V, logical 1 = +8 to +10 V. Maximum output current drive for normal

CMOS gates is about 0.8 mA when operating on a 5 V supply with a short-circuit current of about 4 mA; and about 1.6 mA when operating on a 10 V supply with a short-circuit current of about 11 mA.

The input impedance of CMOS devices is typically 10^{12} Ω with a typical capacitance of 5 pF, and the propagation delay depends upon the fan-out— typically 20 ns for two inputs, which increases by about 5 ns for each 5 pF load. In cases where high speed is not important, the fan-out may be increased to about 50.

Summarizing, CMOS devices offer high packing density, wide supply voltage range, high noise immunity, and low power consumption, all of which, together with reducing costs, have contributed to the increasing applications of CMOS devices.

6.17 CMOS, NOR, and NAND gates

A simple *functional* circuit for a CMOS NOR gate is shown in Fig. 6.13, which uses two *p*-channel MOS devices and two *n*-channel MOS devices.

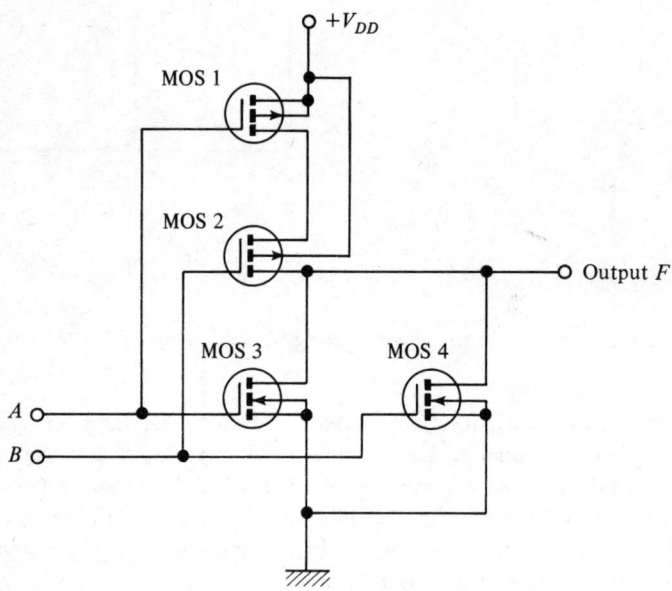

Fig. 6.13. CMOS 2-i/p NOR gate functional circuit.

When both the input signals at A and B are at a logical low level, MOS 3 and MOS 4 (*n* channel) are cut off and MOS 1 and MOS 2 (*p* channel) are switched on, giving a logical high level at output F. When the input signal at A is at a logical low level, and the input signal at B is at a logical high level, MOS 1 and MOS 3 are saturated, and MOS 2 and MOS 4 are cut off, producing a logical low level output at F. This is also the case when A is at a logical high and B is at a logical low level. When both input signals at A and B are at a logical high level, MOS 1 and MOS 2 are cut off, and MOS 3 and MOS 4 are saturated, the output at F is at a logical low level, thus producing the logical NOR function.

The *functional* circuit for a CMOS NAND gate is shown in Fig. 6.14. When both the input signals at A and B are at logical 0, MOS 1 and MOS 2 are switched on and MOS 3 and MOS 4 are cut off, giving a logical 1 level at the output F.

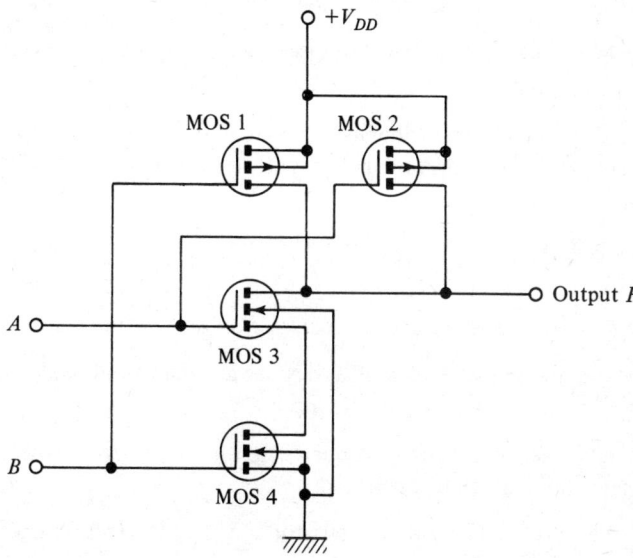

Fig. 6.14. CMOS 2-i/p NAND gate functional circuit.

When A is at logical 0 and B is at logical 1, MOS 1 and MOS 3 are cut off and MOS 2 and MOS 4 are saturated, giving a logical 1 output at F. Similarly, when A is logical 1 and B is logical 0, MOS 1 and MOS 3 are saturated and MOS 2 and MOS 4 are cut off, producing a logical 1 at the output F. When both signals at A and B are at logical 1 levels, MOS 1 and MOS 2 are cut off and MOS 3 and MOS 4 are saturated, producing a logical 0 level at the output F.

EXERCISES TO CHAPTER 6

1. List the factors which affect the choice of logic family for a particular application.

2. With the aid of sketches explain the meaning of the term *propagation delay*, as applied to logic devices.

3. Explain why bipolar junction devices are slower in operation than Schottky junction devices.

4. Explain the meaning of the term *noise immunity* when applied to logic gates.

5. Explain the meaning of the terms *fan-in* and *fan-out*.

6. State the limits of the two common temperature ranges to which IC devices are manufactured, and identify them by name.

7. Sketch the schematic (or functional) circuit of a 2-input TTL NAND gate, and describe its operation.

8. Explain the meaning of the abbreviation TTL.

9. Explain the meaning of the terms totem-pole output, open-collector output, and tri-state output.

10. State the meaning of the following abbreviations and state typical values of power dissipation, propagation delay, and noise immunity: HTTL, LTTL, STTL, LSTTL.

11. With the aid of sketches, explain the meaning of sinking and sourcing currents.

12. State typical voltage levels and currents for all logical states in a TTL gate.

13. Describe the method by which TTL devices are numbered, and explain the meaning of the characters used in the number.

14. Sketch the schematic (or functional) circuit of an open-collector TTL NAND gate, and explain its operation.

15. Explain the purpose of a 'pull-up' resistor and state the factors affecting the choice of resistance value.

16. State a typical application for an open-collector TTL NAND gate and describe its operation.

17. Sketch the schematic (or functional) circuit of an ECL logic gate and describe its operation.

18. Describe the differences between TTL and ECL.

19. List the essential differences between TTL and CMOS. Identify those that are advantages and those that are disadvantages.

20. Sketch the schematic (or functional) circuit of a CMOS NOR gate and describe its operation.

21. Sketch the schematic (or functional) circuit of a CMOS NAND gate and describe its operation.

7 Universal NAND and NOR logic

7.1 NAND and NOR gates

In practice, a particular range of integrated circuits contains only a limited range of gate types. Generally, the restrictions are that only NAND gates, or only NOR gates are used.

It is more economical to use all of one particular type of logic gate in a digital design—a technique which has been developed over the years—so that a particular logic system will generally be made up using all NAND gates *only* or all NOR gates *only*. Although systems may contain NAND gates only or NOR gates only, it must be stressed that the basic logic functions which need to be performed are AND, OR, and NOT.

An examination of the 2-i/p NAND gate together with its truth table, shown in Fig. 7.1, reveals that as long as *one* of the input logical levels is at logical 1, then the *output* logical state is the *complement* of the other input, i.e., the gate becomes a NOT gate, or inverter.

Similarly, an examination of the 2-i/p NOR gate together with its truth table, shown in Fig. 7.2, reveals that as long as *one* of the input logic levels is at logical 0, then the *output* logic state is the *complement* of the other input, i.e., the gate becomes a NOT gate, or inverter.

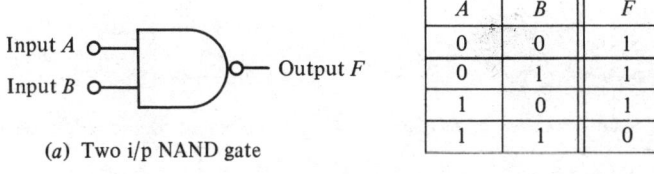

A	B	F
0	0	1
0	1	1
1	0	1
1	1	0

(a) Two i/p NAND gate

(b) Truth table

Fig. 7.1. The NAND logic gate.

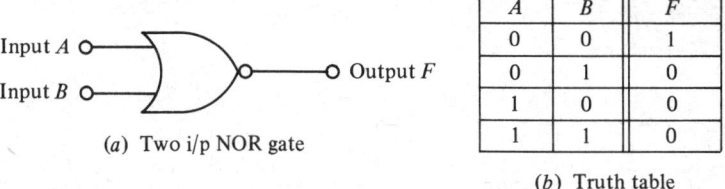

A	B	F
0	0	1
0	1	0
1	0	0
1	1	0

(a) Two i/p NOR gate

(b) Truth table

Fig. 7.2. The NOR logic gate.

7.2 Connection of 'unused' inputs

NAND and NOR logic gates having *several* inputs may be used to effect the NOT gate function (as shown above for 2-i/p gates), *or* this principle may be applied to produce a NAND gate or a NOR gate function having a reduced number of inputs. However, *we must not have any 'floating' inputs*, i.e., all inputs must be connected somewhere whatever logic family is being used. An *unused* input may be connected to a *used* input, but this has the disadvantage that it can increase the loading on the source that provides the logic signal. Alternatively:

(a) *NAND logic*. Unused inputs must be connected to $+V$ through a resistor—to protect the input from transients. For TTL gates, a 1 kΩ resistor is suitable. The same 1 kΩ resistor will suffice for up to 25 unused inputs. In CMOS gates, a 200 kΩ resistor is suitable.

(b) *NOR logic*. Unused inputs must be connected to GND (0 V)—no resistor is required. This applies for both TTL and CMOS gates.

7.3 The logical NOT function

The NOT function, or *inverter*, may therefore be achieved by using NAND gates only, as shown in Fig. 7.3. This principle can be extended to gates having more than two inputs.

NOR gates only may be used to perform the NOT function as shown in Fig. 7.4. Similar principles may be applied to use gates having more than two inputs.

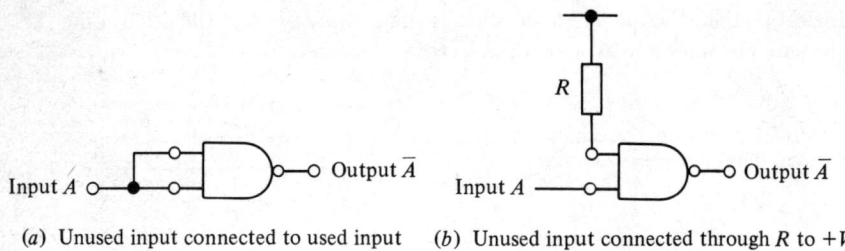

(a) Unused input connected to used input (b) Unused input connected through R to $+V$

TTL: R = 1 kΩ

CMOS: R = 220 kΩ

Fig. 7.3. The NOT logic function using NAND gates only.

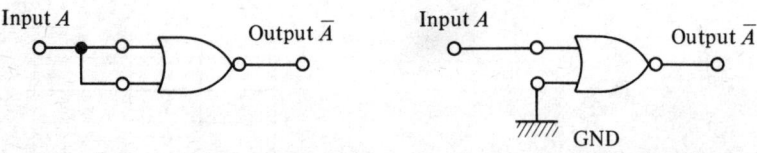

(a) Unused input connected to used input (b) Unused input connected to GND

Fig. 7.4. The NOT logic function using NOR gates only.

Normally, both TTL and CMOS logic families include packages which commonly accommodate *six* (i.e., *hex*) inverters.

7.4 The logical AND function

The basic AND logic function can be performed by using NAND gates only, as shown in Fig. 7.5(*a*), in which the second gate is used as an inverter. Boolean algebra may be used throughout the network to verify the operation of the system.

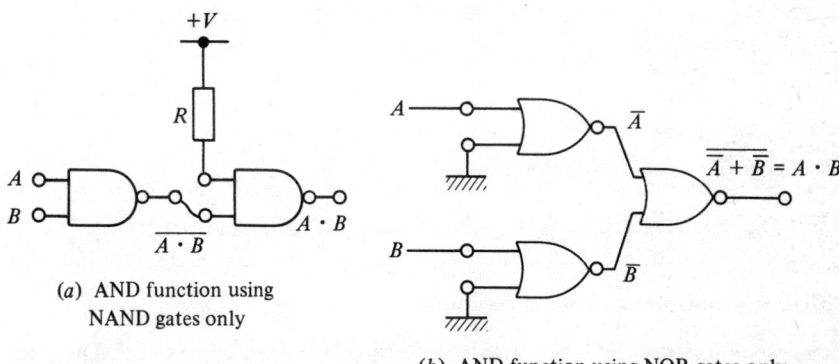

(*a*) AND function using
     NAND gates only

(*b*) AND function using NOR gates only

Fig. 7.5. The AND logic function using NAND gates and NOR gates.

Similarly, the logical AND function can be performed by using NOR gates only, as shown in Fig. 7.5(*b*), in which the two input gates are arranged to operate as inverters. Boolean algebra may be used throughout the network shown, and *De Morgan's dual* must be used to verify that the final output expression is, in fact, the same as the basic expression, for example, $\overline{\overline{A} + \overline{B}} = A.B$.

7.5 The logical OR function

The basic OR logic function can be performed by using NAND gates only, as shown in Fig. 7.6(*a*), in which the two input gates are arranged as inverters. Boolean algebra may be used throughout the network to verify the final expression. In this case, De Morgan's dual must be used to confirm that the final expression at the output is:

$$\overline{\overline{A}.\overline{B}} = A + B$$

Similarly, the logical OR function can be performed by using NOR gates only, as shown in Fig. 7.6(*b*), in which the second gate is arranged as an inverter. Boolean algebra may be used throughout the network to verify the operation of the system.

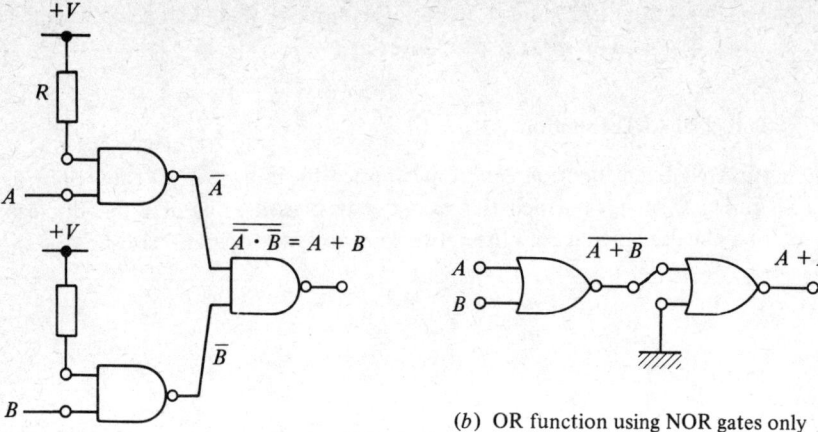

(a) OR function using NAND gates only

(b) OR function using NOR gates only

Fig. 7.6. The OR logic function using NAND gates and NOR gates.

7.6 Logic networks and minimization

Logic networks may now be constructed—using NAND gates only, or NOR gates only—which apply the principles considered in the foregoing paragraphs. One technique is to construct the logic network using the basic logic functions and then to replace each basic logic function by the most suitable equivalent network which uses NAND or NOR networks. An important step is then necessary—to check through the resultant network to eliminate *redundant* gates, i.e., *minimization*. This process is made easier by writing the boolean expressions as they are developed throughout the network.

EXAMPLE 7.1

Construct a logic network which uses NAND gates only to satisfy the logic function F.

$$F = A.B + C.D$$

Deduce the minimal NAND gate network by eliminating redundant gates.

Solution:
Considering *basic* logic gates, the network which satisfies the logic function F is as shown in Fig. 7.7(a).

If each *separate* basic gate is now replaced by its NAND gate equivalent (as shown in Figs 7.3, 7.5(a), and 7.6(a)), then the network becomes as shown in Fig. 7.7(b).

Careful examination of this network (Fig. 7.7(b)) reveals the existence of redundant gates: the equivalent minimal NAND network is shown in Fig. 7.7(c).

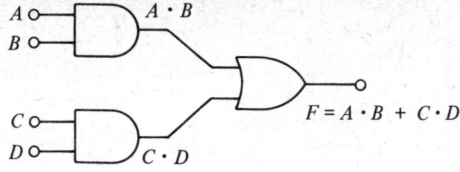

(a) Using basic gates only

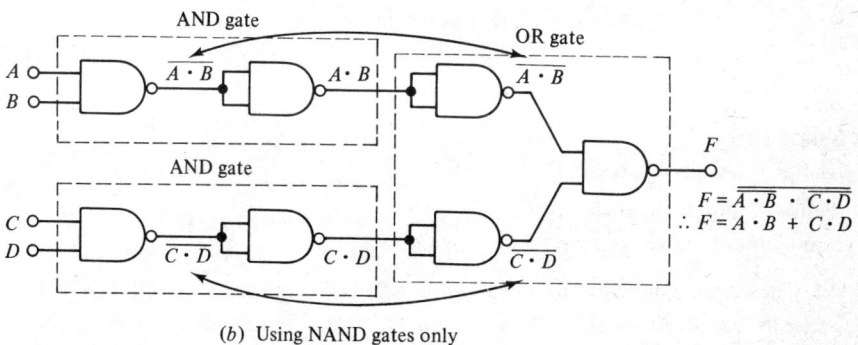

(b) Using NAND gates only

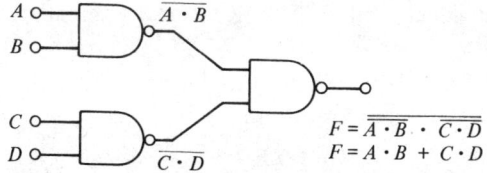

(c) Resultant minimal NAND gate network

Fig. 7.7. NAND gate implementation, and minimization (Example 7.1).

Note: A useful technique when drawing logic networks from boolean expressions is to start from the output F and work backwards, for example, $F = A.B + C.D$ may be rewritten as $F = P + Q$—which is simply a 2-i/p OR gate. Each of the inputs (P and Q) to the OR gate can be represented directly as 2-i/p AND gates, as shown in Fig. 7.7(a).

7.7 Minimization using boolean algebra

The development of the minimal NAND logic network may be simplified by using boolean algebra and, in particular, De Morgan's dual. This technique is applied on the assumption that the boolean expression is already in its simplest form, and the sequence is as follows:

1. Complement the whole expression.
2. Apply De Morgan's dual to produce NAND terms.
3. Complement the whole expression to link terms by NAND and return to the original function.

EXAMPLE 7.2

Produce the minimal NAND expression, and hence minimal NAND logical network, for the boolean expression (as in Example 7.1).

$$F = A.B + C.D$$

Solution:

Complement the expression $\overline{F} = \overline{A.B + C.D}$

Apply De Morgan's dual $\overline{F} = \overline{AB} . \overline{CD}$

Complement the expression $F = \overline{\overline{A.B}.\overline{C.D}}$

The expression thus produced contains NAND terms only, and it can clearly be seen that the logical network which satisfies this expression is exactly the same as that already drawn in Fig. 7.7(c).

The development of minimal NOR logic networks can be achieved by using similar techniques:

1. Complement the whole expression.
2. Rearrange the individual terms into NOR elements, i.e., apply De Morgan's dual to the individual terms.
3. Complement the whole expression.

EXAMPLE 7.3

Deduce the minimal NOR logical network for the function F.

$$F = A + \overline{B}.C$$

Solution:

Complement the expression $\overline{F} = \overline{A + \overline{B}.C}$

Rearrange the terms $\overline{F} = \overline{A} + \overline{B + \overline{C}}$

This expression can be directly implemented using NOR gates only, *but* produces the complement of the required function. Therefore, the logic network using NOR gates only may be drawn as shown in Fig. 7.8, and *complemented* by following with an inverter, as shown.

Note: The application of De Morgan's dual as in Example 7.3 can often appear somewhat obscure. This may be simplified by applying the following rules:

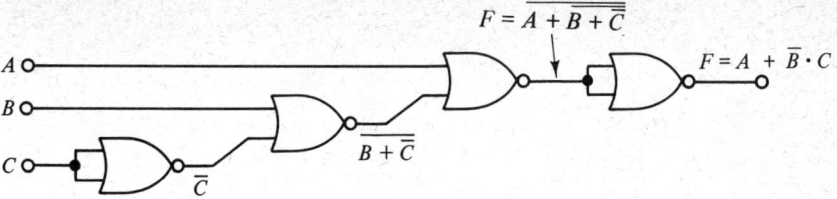

$F = \overline{A + \overline{B + \overline{C}}}$

A

B

C

$\overline{C}$

$\overline{B + \overline{C}}$

$F = A + \overline{B} \cdot C$

Fig. 7.8. Minimal NOR logic network for Example 7.3.

1. Complement the logical variables.
2. Change AND to OR, and OR to AND.
3. Complement the whole expression.

Thus, the term $\overline{B}.C$ in Example 7.3 may be modified:

Complement the logical variables $B.\overline{C}$
Change the linking logic function $B + \overline{C}$
Complement the whole expression $\overline{B + \overline{C}}$

This application illustrates how De Morgan's dual may be applied to only part of a boolean expression.

   In practice, logical systems which are developed using either NAND gates only or NOR gates only in fact use inverters (*not* gates); consequently minimal solutions developed in the ensuing text will assume that inverters are available.

EXAMPLE 7.4

Derive the minimal NAND gate only and the minimal NOR gate only networks for the boolean expression

$$F = A.C + \overline{B}.C$$

Solution:

NAND functions

Complement the original expression $\overline{F} = \overline{A.C + \overline{B}.C}$
Apply De Morgan's dual $\overline{F} = \overline{A.C} . \overline{\overline{B}.C}$
Complement the expression $F = \overline{\overline{A.C} . \overline{\overline{B}.C}}$

   This expression can be directly implemented using NAND gates as shown in Fig. 7.9(*a*).

NOR functions

Complement the original expression $\overline{F} = \overline{A.C + \overline{B}.C}$
Rearrange to NOR terms $\overline{F} = \overline{\overline{A + \overline{C}} + \overline{B + \overline{C}}}$

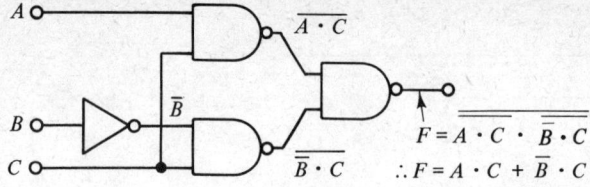

(a) NAND gates only

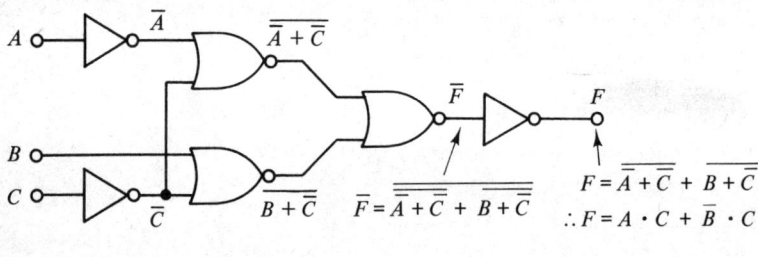

(b) NOR gates only

Fig. 7.9. Minimal networks for Example 7.4.

This expression can be directly implemented using NOR gates only and complemented to reproduce the original function F as shown in Fig. 7.9(b).

EXAMPLE 7.5

Derive the minimal NAND gate only and the minimal NOR gate only networks for the boolean expression

$$F = \bar{A}.\bar{C}.D + B.C.D$$

Solution:

NAND functions

Complement the original expression $\quad \bar{F} = \overline{\bar{A}.\bar{C}.D + B.C.D}$

Apply De Morgan's dual $\qquad\qquad \bar{F} = \overline{\bar{A}.\bar{C}.D} . \overline{B.C.D}$

Complement the expression $\qquad\quad F = \overline{\overline{\bar{A}.\bar{C}.D} . \overline{B.C.D}}$

This expression can be directly implemented using NAND gates as shown in Fig. 7.10(a).

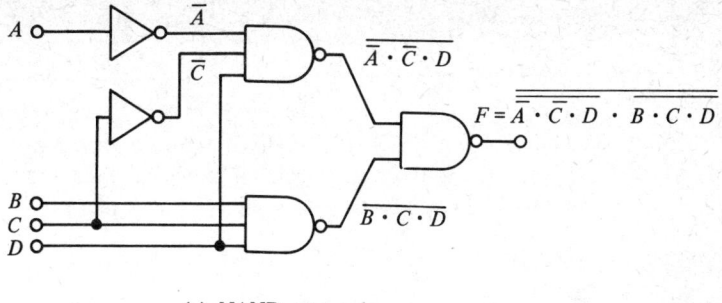

(a) NAND gates only

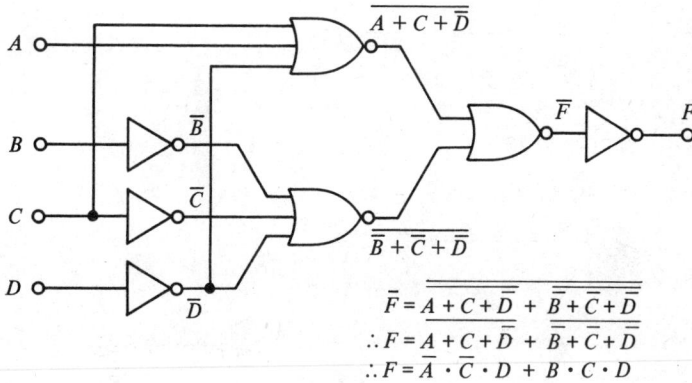

(b) NOR gates only

Fig. 7.10. Minimal networks for Example 7.5.

NOR functions

Complement the original expression $\quad \overline{F} = \overline{\overline{A.\overline{C}.D + B.C.D}}$

Rearrange to NOR terms $\quad \overline{F} = \overline{A + C + \overline{D}} + \overline{\overline{B} + \overline{C} + \overline{D}}$

This expression can be directly implemented using NOR gates only and complemented to reproduce the original function F as shown in Fig. 7.10(b).

7.8   EXCLUSIVE-OR logic function

A very useful function in logic systems is the EXCLUSIVE-OR, or NOT-EQUIVA-LENT, which produces an *output of logical 1* when the two input logic levels are *different* (i.e., logical 1 and logical 0, or logical 0 and logical 1) and gives an *output of logical 0* when the two input logic levels are the *same* (i.e., both logical

0 or both logical 1). Therefore, the boolean equation for the EXCLUSIVE-OR function is

$$F = A.\bar{B} + \bar{A}.B$$

The truth table for the EXCLUSIVE-OR function is shown in Fig. 7.11(a), and the logic network using basic logic gates is shown in Fig. 7.11(b), with the logic symbol for an EXCLUSIVE-OR gate shown in Fig. 7.11(c).

A	B	F
0	0	0
0	1	1
1	0	1
1	1	0

(a) Truth table

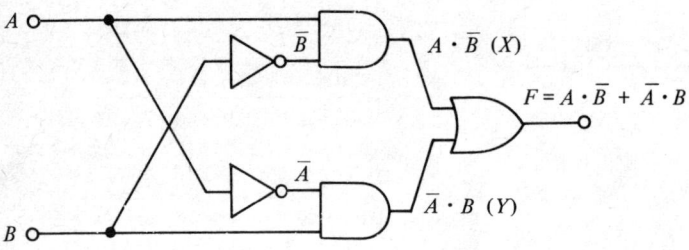

(b) Logic network using basic logic gates

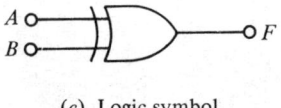

(c) Logic symbol

Fig. 7.11. The EXCLUSIVE-OR logic function.

EXAMPLE 7.6

Derive the minimal NAND gate only network for the EXCLUSIVE-OR function.

$$F = A.\bar{B} + \bar{A}.B$$

Solution:

Complement the original expression $\quad \overline{F} = \overline{A.\overline{B} + \overline{A}.B}$

Apply De Morgan's dual $\qquad\qquad \overline{F} = \overline{A.\overline{B}} \cdot \overline{\overline{A}.B}$

Complement the whole expression $\quad F = \overline{\overline{A.\overline{B}} \cdot \overline{\overline{A}.B}}$

This expression may be directly implemented using NAND gates only, as shown in Fig. 7.12. However, examination of this network shows a *possible* simplification. The upper NAND gate requires a $\overline{B}$ input, and the lower NAND gate requires an $\overline{A}$ input. In many cases, the logic levels required as individual signals may sometimes be combined into *one* NAND gate, the output of which is applied to *both* upper and lower NAND gates. When this is applied in this case, the logic network shown in Fig. 7.13 is produced—which does, in fact, perform the EXCLUSIVE-OR function as shown by the boolean algebra in Fig. 7.13.

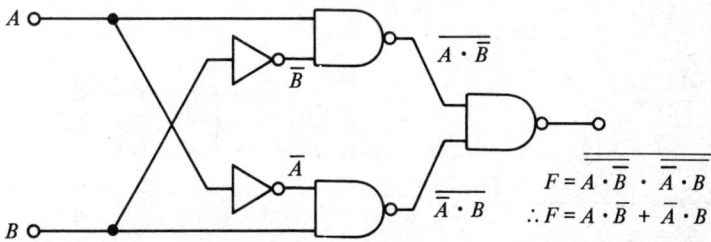

Fig. 7.12 NAND network for EXCLUSIVE-OR (Example 7.6).

EXAMPLE 7.7

Derive the minimal NOR gate only network for the EXCLUSIVE-OR function

$$F = A.\overline{B} + \overline{A}.B$$

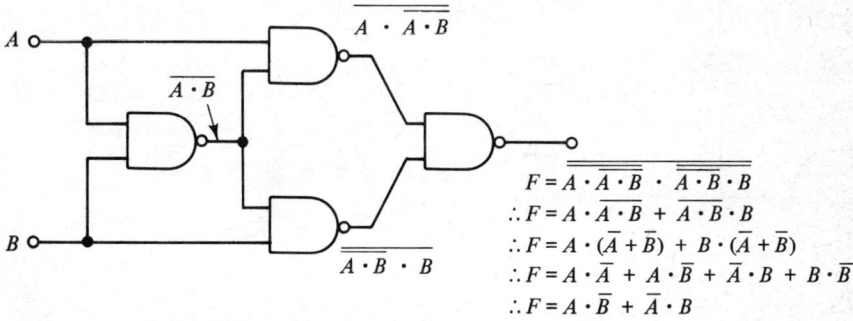

Fig. 7.13. Minimal NAND gate only network for EXCLUSIVE-OR function
(Example 7.6).

Solution:

Complement the original expression $\overline{F} = A.\overline{B} + \overline{A}.B$
Rearrange to NOR terms,
    i.e., apply De Morgan's dual to
    the individual terms $\overline{F} = \overline{\overline{A} + B} + \overline{A + \overline{B}}$

This expression may be directly implemented using NOR gates only, and then complemented to reproduce the original function F as shown in Fig. 7.14. However, examination of this network reveals a *possible* simplification (similar to Example 7.6). The resulting logic network shown in Fig. 7.15 can be seen to perform the EXCLUSIVE-OR function, and is therefore the minimal NOR gate only network.

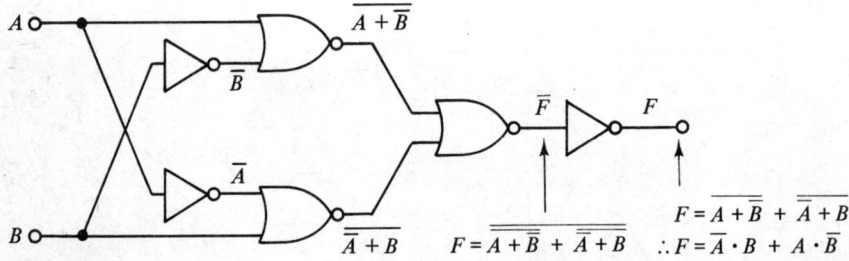

Fig. 7.14. NOR network for EXCLUSIVE-OR (Example 7.7).

7.9 Logic comparator

A very useful arrangement is the logic comparator, which allows two logic signals to be compared. If $A = 1$ and $B = 0$, then $A > B$, and if $A = 0$ and $B = 1$ then

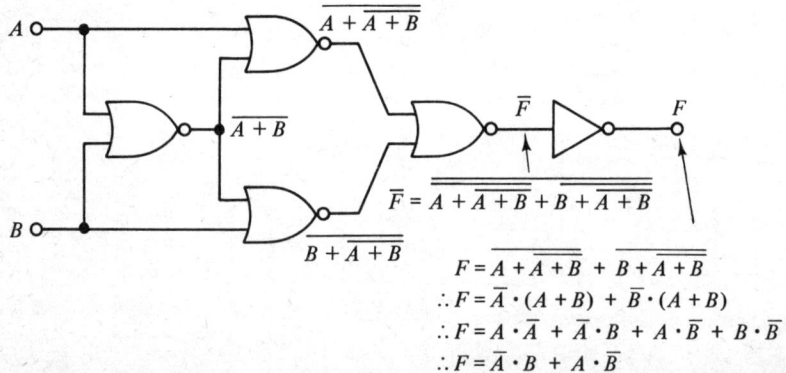

Fig. 7.15. Minimal NOR gate only network for EXCL-OR function (Example 7.7).

$A < B$. Examination of the logic network for the EXCLUSIVE-OR function shown in Fig. 7.11(b) reveals that point X corresponds to $A > B$ and will only produce a logical 1 at X for this condition. Similarly, the output at Y corresponds to $A < B$, that is, Y will only be at logical 1 for the condition $A < B$. ($A = 0$, $B = 1$.)

If *corresponding* positions are used in the NAND networks shown in Figs 7.12 and 7.13 a *similar* result is produced. However, taking the two outputs X and Y from the outputs of the upper and lower NAND gates reveals that these two signals are logical 1 for all conditions except $A > B$ and $A < B$ respectively. Therefore, it is necessary to use inverters from these outputs for the NAND network as shown in Fig. 7.16(a), which represents the logical comparator. The truth table for the comparator is shown in Fig. 7.16(b).

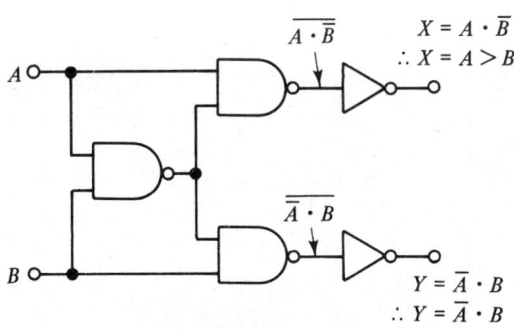

$$X = A \cdot \overline{B}$$
$$\therefore X = A > B$$

$$Y = \overline{A} \cdot B$$
$$\therefore Y = \overline{A} \cdot B$$

Inputs		Outputs	
A	B	X	Y
0	0	0	0
0	1	0	1
1	0	1	0
1	1	0	0

(a) Logical network   (b) Truth table

Fig. 7.16. The logical comparator (NAND gates only).

7.10 Simple machine safety interlock system

Consider a simple boring machine which is driven by an electric motor. The motor must operate *F only* when the *power supply switch S* is operated *and* the following safety features are satisfied:

1. A *safety guard G* is in position.
2. The motor overload *current limit L* is in a safe state.

In addition to the above, *maintenance facilities* must be provided such that when a *maintenance key K* is inserted, the motor may be run *without* the safety guard in position, while all other requirements are as specified.

Assume that the logic signals for each of these requirements are obtained from transducers which are arranged to produce a logical 1 level when the variables are in the *safe* operating state and a logical 0 for all other conditions.

Examination of the above requirements reveals that the system has *four* inputs S, G, L, and K. The conditions described can be represented in a truth table as shown in Fig. 7.17.

From the truth table, it can be seen that the motor operates ($F = 1$), that is,

S	G	L	K	F
0	0	0	0	0
0	0	0	1	0
0	0	1	0	0
0	0	1	1	0
0	1	0	0	0
0	1	0	1	0
0	1	1	0	0
0	1	1	1	0
1	0	0	0	0
1	0	0	1	0
1	0	1	0	0
1	0	1	1	1
1	1	0	0	0
1	1	0	1	0
1	1	1	0	1
1	1	1	1	1

Fig. 7.17. Truth table for safety interlock system.

the function F is realized for *three* different combinations of the four input signals so that the boolean equation describing the function F may be written from the truth table

$$F = S.\bar{G}.L.K + S.G.L.\bar{K} + S.G.L.K \qquad (7.1)$$

This boolean equation (7.1) describes the function F in terms of the *basic* logic gates for which the logic network is as shown in Fig. 7.18.

Now it is generally required to use a *minimum* number of the *available* logic gates, i.e., usually NAND only or NOR only. To derive the minimal NAND network (or NOR network) requires that the boolean equation describing the function is reduced to its simplest form

$$F = S.\bar{G}.L.K + S.G.L.\bar{K} + S.G.L.K$$
$$\therefore F = S.\bar{G}.L.K + S.G.L.K + S.G.L.\bar{K} + S.G.L.K$$
$$\therefore F = S.L.K (\bar{G} + G) + S.G.L (\bar{K} + K)$$
$$\therefore F = S.L.K + S.G.L \qquad (7.2)$$

or

$$F = S.L (K + G) \qquad (7.3)$$

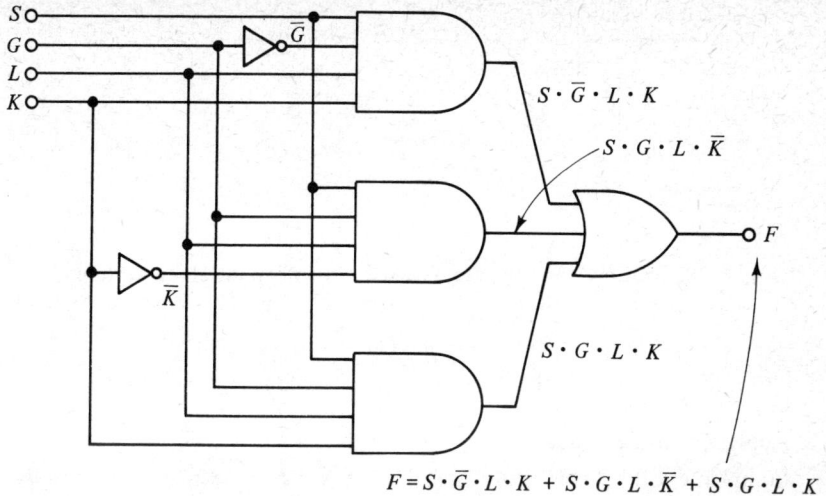

$$F = S \cdot \overline{G} \cdot L \cdot K + S \cdot G \cdot L \cdot \overline{K} + S \cdot G \cdot L \cdot K$$

Fig. 7.18. Safety interlock system using basic logic gates.

Equation (7.3) represents a simplification of the boolean equation (7.1) for this function, and the logic network using *basic* gates is shown in Fig. 7.19, which shows that a reduced number of gates have been used when compared with the network shown in Fig. 7.18.

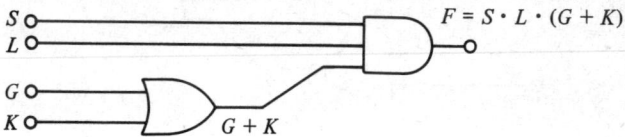

Fig. 7.19. Simplified basic gate network.

To derive the minimal NAND gate expression and logic network, Eq. (7.2) is more suitable (for NAND) than Eq. (7.3). This may be verified by attempting to produce the minimal NAND network from Eq. (7.3) and comparing the result with the following solution

$$F = S.G.K + S.G.L$$

Complement the original expression $\quad \overline{F} = \overline{S.G.K + S.G.L}$

Apply De Morgan's dual $\qquad\qquad \overline{F} = \overline{S.G.K} \cdot \overline{S.G.L}$

Complement the whole expression $\quad F = \overline{\overline{S.G.K} \cdot \overline{S.G.L}}$ $\qquad$ (7.4)

Equation (7.4) may be directly implemented using NAND gates only to produce the minimal NAND gate network shown in Fig. 7.20.

Equation (7.3) is the most suitable form of original expression to use to derive the minimal NOR gate network. This may be verified by attempting to produce the minimal NOR network from Eq. (7.2), and comparing the result with the following solution

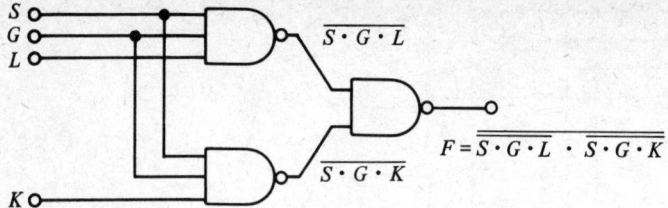

Fig. 7.20. Minimal NAND gate network for safety interlock system.

$$F = S.L.\,(K + G)$$

Complement the original expression $\quad \overline{F} = \overline{S.L.\,(K + G)}$

Apply De Morgan's dual $\qquad\qquad \overline{F} = \overline{S} + \overline{L} + \overline{K} + G$

Complement the whole expression $\quad F = \overline{\overline{S} + \overline{L} + \overline{K} + G}$ $\qquad$ (7.5)

Equation (7.5) can be directly implemented using NOR gates only to produce the minimal NOR gate network shown in Fig. 7.21.

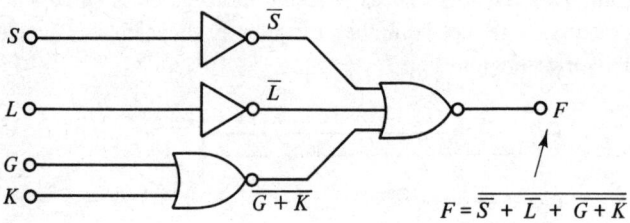

Fig. 7.21. Minimal NOR gate network for safety interlock system.

EXERCISES TO CHAPTER 7

1. Explain why most industrial logic systems use either NAND gates only or NOR gates only.

2. With the aid of sketches show how a 2-input NAND gate may be used as an *inverter* (NOT gate).

3. With the aid of a truth table, describe how a 2-input NOR gate may be used as an *inverter*.

4. With the aid of sketches describe in detail the precautions which must be observed relating to the connection of *unused inputs* on NAND and NOR logic gates.

5. With the aid of sketches, show how NAND gates only may be connected to perform the AND, OR, and NOT logic functions.

6. Using boolean algebra describe how NOR gates only may be connected to perform the AND, OR, and NOT logic functions.

7. Explain the meaning of the term *minimization.*

8. Briefly describe a technique for the minimization of a logic system in which a solution in NAND terms is required.

9. Derive the minimal NAND logic network for the boolean equation

$$F = A.C + \bar{B}.C$$

10. Derive the minimal NOR logic network for the boolean equation

$$F = A.B.C + A.\bar{B}.\bar{C}$$

11. Simplify the following boolean equation

$$F + A.\bar{C} + B.\bar{C}.\bar{D} + A.B.\bar{D}$$

12. Derive the minimal NOR gate network for the boolean equation

$$F = \bar{A}.\bar{B}.D + \bar{A}.C + \bar{B}.C.D$$

13. Simplify the following boolean equation

$$F = \bar{A}.\bar{B}.\bar{C}.\bar{D} + \bar{A}.\bar{B}.\bar{C}.D + A.\bar{B}.\bar{C}.\bar{D} + A.\bar{B}.\bar{C}.D$$

Hence, derive the minimal NAND gate only network.

14. Sketch a logical network using NAND gates only which will perform the function of the EXCLUSIVE-OR function.

15. Draw the truth table for the EXCLUSIVE-OR function.

16. Explain the meaning of the term *comparator* and sketch a logic network using NAND gates only which will perform the comparator function.

17. With the aid of sketches describe how NAND only or NOR only logic gates may be used to solve an identical logic control problem.

8 Karnaugh maps

8.1 Introduction

The *Karnaugh* map is a compact graphical method of representing the truth table for a given logic system. It is a rectangular diagram, the area of which is divided into *cells*, where

$$\text{Total number of cells in the map} = 2^N$$

where N = number of logical variables in the system.

Each logical variable in a Karnaugh map is represented by *half* the total area and *its complement is represented by the other half.*

Karnaugh maps for *two*- and *three*-variable systems are as shown in Figs 8.1(a) and (b) respectively, in which the boolean expression (represented by that cell, for that combination of the logical variables) has been written *inside* each cell—derived in similar fashion to that of identifying the squares in the game 'Battleships'.

The Karnaugh map for a *four*-variable system is shown in Fig. 8.2(a). Again, the boolean expression for each combination of the four variables is written inside each cell.

If we now use a logical 1 to represent each of the variables A, B, C, and D, and a logical 0 to represent the complements, that is, $\bar{A}, \bar{B}, \bar{C}$, and $\bar{D}$, and redraw the Karnaugh map shown in Fig. 8.2(a) to that shown in Fig. 8.2(b), the resulting diagram should strictly be called a *Veitch* diagram. However, to avoid the possibility of confusion, we will refer to all these diagrams as Karnaugh maps.

In the diagram shown in Fig. 8.2(b), the cells within the map have been completed using the same techniques as before, except that this time the logic levels of 1 and 0 have been used, producing an array of 1s and 0s which appears to be similar to a 4-bit binary number. The reader is advised to confirm that the two diagrams shown in Fig. 8.2 are in fact different methods of representing the same truth table.

	$\bar{B}$	B
$\bar{A}$	$\bar{A} \cdot \bar{B}$	$\bar{A} \cdot B$
A	$A \cdot \bar{B}$	$A \cdot B$

	$\bar{B}$	$\bar{B}$	B	B
$\bar{A}$	$\bar{A} \cdot \bar{B} \cdot \bar{C}$	$\bar{A} \cdot \bar{B} \cdot C$	$\bar{A} \cdot B \cdot C$	$\bar{A} \cdot B \cdot \bar{C}$
A	$A \cdot \bar{B} \cdot \bar{C}$	$A \cdot \bar{B} \cdot C$	$A \cdot B \cdot C$	$A \cdot B \cdot \bar{C}$

| | $\bar{C}$ | C | C | $\bar{C}$ |

(a) Two variables = 2^2 = 4 cells   (b) Three variables = 2^3 = 8 cells

Fig. 8.1. Karnaugh maps for two- and three- variable systems.

	$\bar{C}$	$\bar{C}$	C	C	
$\bar{A}$	$\bar{A}\cdot\bar{B}\cdot\bar{C}\cdot\bar{D}$	$\bar{A}\cdot\bar{B}\cdot\bar{C}\cdot D$	$\bar{A}\cdot\bar{B}\cdot C\cdot D$	$\bar{A}\cdot\bar{B}\cdot C\cdot\bar{D}$	$\bar{B}$
$\bar{A}$	$\bar{A}\cdot B\cdot\bar{C}\cdot\bar{D}$	$\bar{A}\cdot B\cdot\bar{C}\cdot D$	$\bar{A}\cdot B\cdot C\cdot D$	$\bar{A}\cdot B\cdot C\cdot\bar{D}$	B
A	$A\cdot B\cdot\bar{C}\cdot\bar{D}$	$A\cdot B\cdot\bar{C}\cdot D$	$A\cdot B\cdot C\cdot D$	$A\cdot B\cdot C\cdot\bar{D}$	B
A	$A\cdot\bar{B}\cdot\bar{C}\cdot\bar{D}$	$A\cdot\bar{B}\cdot\bar{C}\cdot D$	$A\cdot\bar{B}\cdot C\cdot D$	$A\cdot\bar{B}\cdot C\cdot\bar{D}$	$\bar{B}$
	$\bar{D}$	D	D	$\bar{D}$	

(a) Karnaugh map

AB \ CD	00	01	11	10
00	0000	0001	0011	0010
01	0100	0101	0111	0110
11	1100	1101	1111	1110
10	1000	1001	1011	1010

(b) Veitch diagram

Fig. 8.2. Karnaugh maps for four-variable system.

If the *positional weights* of the digits in the cells in Fig. 8.2(b) are ignored, then it can be observed that *adjacent cells* in a Karnaugh map *differ by only one binary digit*. In this sense, cells at the top are adjacent to cells at the bottom, and cells on the right are adjacent to cells on the left. Confirm that this is true before proceeding.

8.2 Function mapping

Mapping is a graphical method of representing logical equations. This technique is widely used to prove boolean theorems, to design logic networks and to assist in the minimization of logic gates in logic function networks.

8.3 Cell looping

This very important principle is best described with the aid of an example.

EXAMPLE 8.1

Suppose that a particular logic function is described by the boolean equation

$$F = A.B.C.D + \bar{A}.B.C.D + \bar{A}.B.\bar{C}.D + \bar{A}.\bar{B}.C.D$$

Use a Karnaugh map to simplify this equation.

Solution:
The truth table for this equation is as shown in Fig. 8.3, in which F (the function) is shown as a logical 1 for each of the conditions described by each of the terms in the boolean equation above, and logical 0 for all other conditions.

The Karnaugh map may be drawn for this system, either directly from the original boolean equation, or from the truth table, *but note* that we now only write a logical 1 in each cell which describes *one* four-variable term in the original equation, and logical 0 in all other cells, as shown in Fig. 8.4(a).

A	B	C	D	F
0	0	0	0	0
0	0	0	1	1
0	0	1	0	0
0	0	1	1	0
0	1	0	0	0
0	1	0	1	1
0	1	1	0	0
0	1	1	1	1
1	0	0	0	0
1	0	0	1	0
1	0	1	0	0
1	0	1	1	0
1	1	0	0	0
1	1	0	1	0
1	1	1	0	0
1	1	1	1	1

Fig. 8.3. Truth table for Example 8.1.

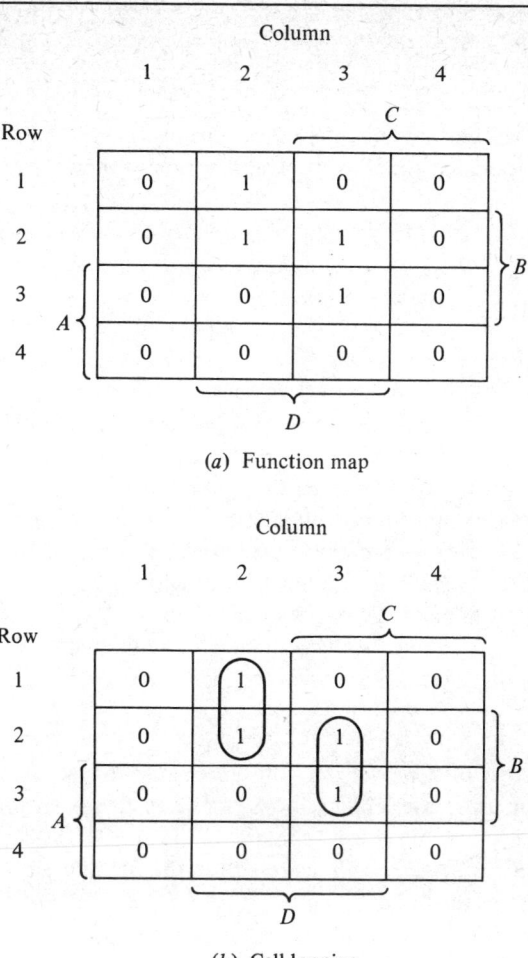

(a) Function map

(b) Cell looping

Fig. 8.4. Karnaugh maps for Example 8.1.

Now consider the *adjacent cells* in rows 1 and 2 of column 2. These two cells correspond to the terms $\overline{A}.B.\overline{C}.D$ and $\overline{A}.\overline{B}.\overline{C}.D$ in the above boolean equation. These two terms may be simplified using boolean algebra techniques

$$\overline{A}.B.\overline{C}.D + \overline{A}.\overline{B}.\overline{C}.D = \overline{A}.\overline{C}.D\,(B + \overline{B})$$
$$= \overline{A}.\overline{C}.D$$

Similarly, consider the adjacent cells in rows 2 and 3 of column 3, which correspond to the terms $A.B.C.D$ and $\overline{A}.B.C.D$. These two terms may also be simplified

$$A.B.C.D + \overline{A}.B.C.D = B.C.D\,(A + \overline{A})$$
$$= B.C.D$$

Thus, the equation describing the logic function F may be rewritten as

$$F = \overline{A}.\overline{C}.D + B.C.D$$

Returning to the Karnaugh map shown in Fig. 8.4(a), pairs of adjacent cells which identify possible simplifications are normally shown looped together, as shown in Fig. 8.4(b).

Note: This looping of cells may be extended to include 4 cells, 8 cells, etc., and we generally try to accommodate the *largest possible number* of cells in a loop. Loops may also overlap one another.

When *two* cells are looped, the corresponding *two* terms in the boolean equation are *combined into a single term* (*which excludes the variable that is in a different logical state* between the two cells).

When *four* cells are looped, the corresponding *four* terms in the boolean equation are *combined into a single term* (*which excludes the two variables that are in different logical states* between the four cells).

When *eight* cells are looped, the corresponding *eight* terms in the boolean equation are *combined into a single term* (*which excludes the three variables that are in different logical states* between the eight cells).

Therefore, examination of the Karnaugh map in Fig. 8.4(b) reveals that, in Example 8.1, we can immediately write down a simplified equation for F which contains two terms only, i.e., one term for each loop

$$F = \overline{A}.\overline{C}.D + B.C.D$$

The Karnaugh map provides a convenient method of simplifying boolean equations. Even using overlapping loops does not always envelope all the cells representing the function, i.e., cells containing a logical 1. The simplified solution *must include all the original cells* representing the function; it is better *included in a loop*, the larger the better.

EXAMPLE 8.2

Draw the Karnaugh map for the following boolean equation

$$F = A.\overline{B}.\overline{C} + A.B.\overline{C}$$

Solution:
Examination of the boolean equation reveals that this is a *three*-variable logic system, and the Karnaugh map is as shown in Fig. 8.5.

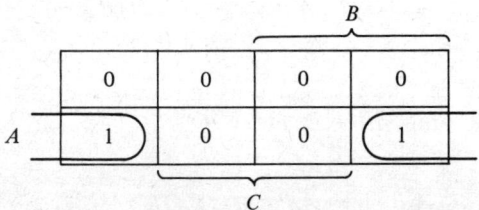

Fig. 8.5. Karnaugh map for Example 8.2.

Careful examination of the *two* cells representing this function in the map reveals that they are, in fact, adjacent to each other and may therefore be looped together, as shown. From the *one* loop shown

$$F = A.\overline{C}$$

EXAMPLE 8.3

Draw the Karnaugh map for the function F described by the boolean equation

$$F = A.\overline{B}.\overline{C}.D + A.\overline{B}.\overline{C}.\overline{D} + \overline{A}.B.C.D + \overline{A}.B.\overline{C}.D$$

Use cell-looping techniques to simplify this function.

Solution:
The Karnaugh map for this four-variable system is shown in Fig. 8.6.
From the *two* loops shown

$$F = A.\overline{B}.\overline{C} + \overline{A}.B.D$$

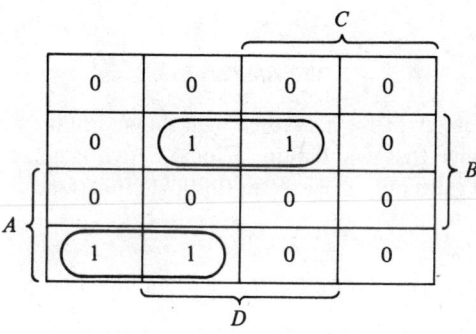

Fig. 8.6. Karnaugh map Example 8.3.

EXAMPLE 8.4

Draw the Karnaugh map for the function F described by the boolean equation

$$F = A.\overline{B}.\overline{C}.\overline{D} + A.\overline{B}.C.\overline{D} + \overline{A}.\overline{B}.C.\overline{D} + \overline{A}.\overline{B}.\overline{C}.\overline{D}$$

Use cell-looping techniques to simplify this function.

Solution:
The Karnaugh map for this system is shown in Fig. 8.7.
From the *one* loop shown

$$F = \overline{B}.\overline{D}$$

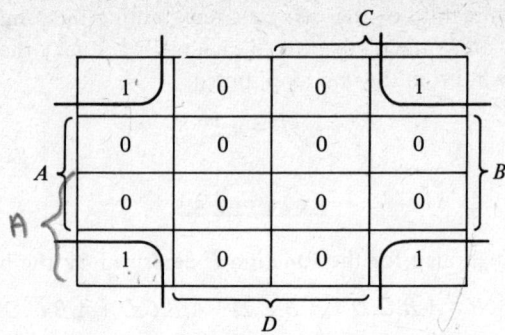

Fig. 8.7. Karnaugh map for Example 8.4.

The Karnaugh maps shown above are really compact truth tables which describe the logic function F using the basic logic operations AND, OR, and NOT. In practice, however, most systems use only NAND gates or only NOR gates, so that we need to examine the applications of the Karnaugh map to meet these requirements.

EXAMPLE 8.5

Suppose that we are given the Karnaugh map shown in Fig. 8.8(a) describing the function F. Assume that we require a logic network comprising a minimum number of NAND gates only to perform this logic function.

Solution:
From the *two* overlapping loops shown, the simplified boolean equation describing this function is

$$F = A.D + A.\overline{C}$$

which describes the function using AND, OR, and NOT operations.

Techniques similar to those described in Chapter 7 must be applied to derive the NAND terms:

1. Complement the simplified boolean equation $\quad \overline{F} = \overline{A.D + A.\overline{C}}$
2. Apply De Morgan's dual $\quad\quad\quad\quad\quad\quad\quad \overline{F} = \overline{A.D} \cdot \overline{A.\overline{C}}$
3. Complement the whole expression $\quad\quad\quad F = \overline{\overline{A.D} \cdot \overline{A.\overline{C}}}$

This equation now contains terms which may be directly implemented using NAND gates only, as shown in Fig. 8.8(b).

EXAMPLE 8.6

Consider the *same* logic function as in Example 8.5. The Karnaugh map is redrawn in Fig. 8.9(a).

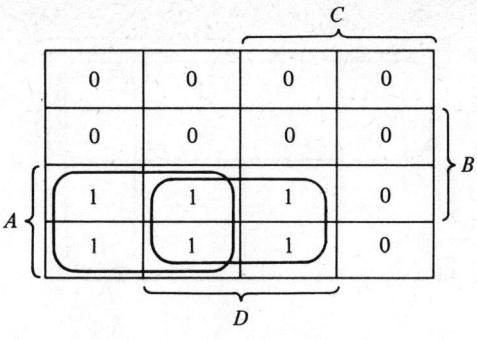

(a) Karnaugh map

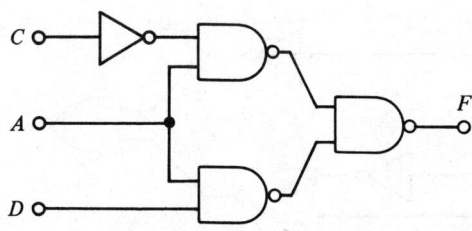

(b) Minimal NAND gate only network

Fig. 8.8 Karnaugh map and logic network for Example 8.5.

Assume that we require a logic network comprising a minimum number of NOR gates only to perform this function.

Solution:
It is more convenient, when deriving a NOR network solution, to loop cells containing logical 0s, which gives a simplified expression for the *complement* of the original function.

Thus, from the *two* overlapping loops shown in Fig. 8.9(a), the boolean equation describing the *complement of the function* is:

$$\overline{F} = \overline{A} + C.\overline{D}$$

which is a simplified equation describing the complement of the function in terms of AND, OR, and NOT gates.

To convert this simplified equation to one which can be directly implemented using NOR gates only:

1. Apply De Morgan's dual to individual terms to produce NOR terms

$$\overline{F} = \overline{A} + \overline{\overline{C} + D}$$

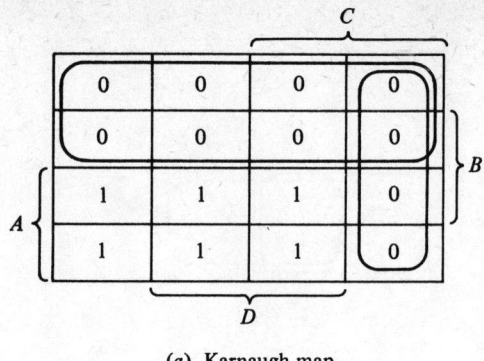

(a) Karnaugh map

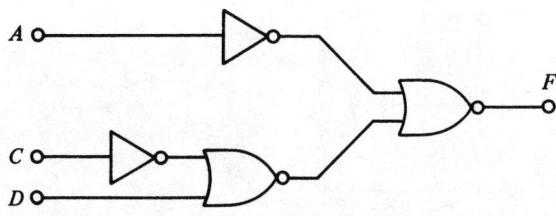

(b) Minimal NOR gate only network

Fig. 8.9. Karnaugh map and logic network for Example 8.6.

2. Complement the whole expression

$$F = \overline{\overline{A} + \overline{\overline{C} + D}}$$

This equation now contains terms which may be directly implemented using NOR gates only as shown in Fig. 8.9(b).

Some situations occur in practice in which a cell (or group of cells) may contain *either* a logical 0 or logical 1. In the Karnaugh map, it is advisable to write an *asterisk* (∗) in such cells, so that when we examine the map for possible loops, we can *include* or *exclude* these cells as required to complete loops of a larger number of cells.

EXAMPLE 8.7

Consider a logic system which has *four* signal inputs designated A, B, C, and D. It may be assumed that the logic levels at A and B represent a binary number X, in which A is the most significant digit (MSD). Similarly, the logic levels at C and D represent a second binary number Y, in which C is the MSD.

The conditions of this system for which the output F is a logical 1 are satisfied when the binary number X is greater than the binary number Y. The output at F is logical 0 for all other values of X and Y.

Draw the Karnaugh map for this logic system, and use cell-looping techniques to derive a simplified boolean equation to describe the function F.

Derive a logical network using a minimum number of NAND gates only which is capable of performing this function.

Solution:

(a) In this case it is advisable to draw up the truth table first, as shown in Fig. 8.10.

The Karnaugh map may now be easily constructed from the truth table as shown in Fig. 8.11(a).

Since, in this case, we are ultimately required to derive a minimal NAND gate network, consider the cells containing logical 1s.

From the *three* overlapping loops shown in the Karnaugh map, the *simplified* boolean equation is

$$F = A.\overline{C} + B.\overline{C}.\overline{D} + A.B.\overline{D}$$

X		Y		F
A	B	C	D	
0	0	0	0	0
0	0	0	1	0
0	0	1	0	0
0	0	1	1	0
0	1	0	0	1
0	1	0	1	0
0	1	1	0	0
0	1	1	1	0
1	0	0	0	1
1	0	0	1	1
1	0	1	0	0
1	0	1	1	0
1	1	0	0	1
1	1	0	1	1
1	1	1	0	1
1	1	1	1	0

Fig. 8.10. Truth table for Example 8.7.

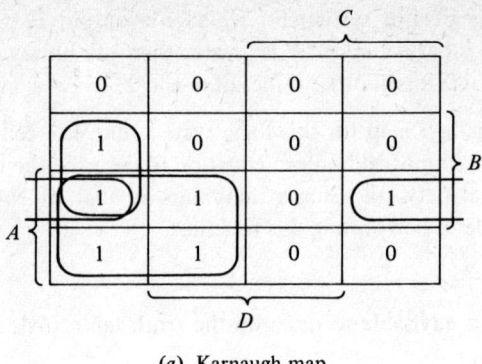

(a) Karnaugh map

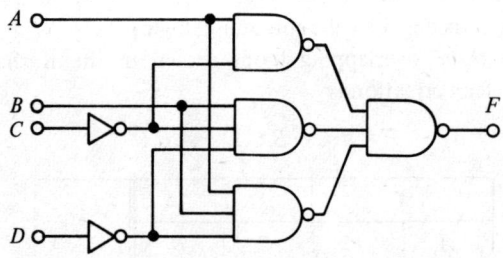

(b) Minimal NAND gate only logical network

Fig. 8.11. Karnaugh map and logic network for Example 8.7.

(b) The above equation describes the function F in terms of the *basic* logic gates. To derive this function in terms of NAND gates:

  (i) complement the simplified boolean equation

  $$\overline{F} = A.\overline{C} + B.\overline{C}.D + A.B.\overline{D}$$

  (ii) apply De Morgan's dual

  $$\overline{F} = \overline{A.\overline{C}} \cdot \overline{B.\overline{C}.D} \cdot \overline{A.B.\overline{D}}$$

  (iii) complement the whole expression

  $$F = \overline{\overline{A.\overline{C}} \cdot \overline{B.\overline{C}.D} \cdot \overline{A.B.\overline{D}}}$$

This equation can now be directly implemented using NAND gates only as shown in Fig. 8.11(b).

EXAMPLE 8.8

The four inputs A, B, C, and D to a logic system represent a four-bit binary number in which A is the most significant digit. If the input is less than or equal to

8_{10}, the output function F is logical 1. When the input is greater than 11_{10}, the output function F may be *either* logical 1 *or* logical 0.

By the use of Karnaugh mapping derive a simple boolean equation which describes this function.

Solution:

The truth table is drawn up as shown in Fig. 8.12. The Karnaugh map may easily be drawn from the truth table as shown in Fig. 8.13. It should be recalled that we may *include* or *exclude* cells containing asterisks in our loops. In this case, it is helpful to *include one* and *exclude three* asterisks to obtain one loop of *eight* cells and one loop of *four* cells.

From the *two* overlapping loops shown in the Karnaugh map

$$F = \overline{A} + \overline{C}.\overline{D}$$

As an exercise in the techniques described, you may now try to derive logic networks using NAND gates only and NOR gates only for the system described in Example 8.8.

Note: It is possible to construct the Karnaugh maps directly from the descriptions of the logic systems in Examples 8.7 and 8.8; the truth tables have been constructed only to simplify the construction of the Karnaugh maps.

A	B	C	D	F
0	0	0	0	1
0	0	0	1	1
0	0	1	0	1
0	0	1	1	1
0	1	0	0	1
0	1	0	1	1
0	1	1	0	1
0	1	1	1	1
1	0	0	0	1
1	0	0	1	0
1	0	1	0	0
1	0	1	1	0
1	1	0	0	*
1	1	0	1	*
1	1	1	0	*
1	1	1	1	*

Fig. 8.12. Truth table for Example 8.8.

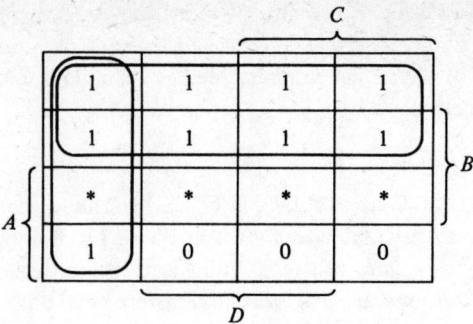

Fig. 8.13. Karnaugh map for Example 8.8.

EXERCISES TO CHAPTER 8

1. Explain how the Karnaugh map can be used to represent the truth table of a logic system.

2. Describe the principle of looping adjacent cells in a Karnaugh map to simplify the boolean equation describing the logic function.

3. Draw the Karnaugh map for the boolean equation

$$F = \overline{A}.\overline{B}.\overline{C}.\overline{D} + \overline{A}.\overline{B}.C.\overline{D}$$

Hence, use cell-looping techniques to simplify the boolean equation.

4. Draw the Karnaugh map for the boolean equation

$$F = \overline{A}.\overline{B}.\overline{C} + A.\overline{B}.\overline{C} + A.\overline{B}.C + \overline{A}.B.\overline{C}$$

Hence, use cell-looping techniques to simplify the boolean equation.

5. Draw the Karnaugh map for the boolean equation

$$F = \overline{A}.B.C + \overline{A}.B.\overline{C} + A.\overline{B}.C + A.\overline{B}.\overline{C}$$

Hence, use cell-looping techniques to simplify the boolean equation.

6. Briefly explain how the Karnaugh map can be used to aid derivation of a logic network which uses NAND gates only.

7. Briefly describe how the Karnaugh map can be used to aid the derivation of a logic network which contains NOR gates only.

8. Using Karnaugh map techniques, derive a minimal NAND gate only logic network to perform the function

$$F = \overline{A}.B.\overline{C} + A.B.\overline{C} + A.\overline{B}.\overline{C}$$

9. Use a Karnaugh map and cell-looping techniques to simplify the boolean equation

$$F = A.\bar{B}.C.D + A.B.C.D + A.\bar{B}.\bar{C}.D + A.B.\bar{C}.D$$

Hence, derive a minimal NAND gate only logic network for this function.

10. Using a Karnaugh map and cell-looping techniques, simplify

$$F = A.B.\bar{C}.\bar{D} + \bar{A}.B.\bar{C}.\bar{D} + \bar{A}.B.C.\bar{D} + A.B.C.\bar{D}$$

11. Use a Karnaugh map and cell-looping techniques, derive a minimal NOR gate only logical network for

$$F = \bar{A}.\bar{B}.\bar{C}.\bar{D} + \bar{A}.\bar{B}.\bar{C}.D + A.\bar{B}.\bar{C}.\bar{D} + A.\bar{B}.\bar{C}.D$$

12. A logic system has four inputs A, B, C, and D, in which A and B represent a binary number X with A the MSB, and C and D represent a binary number Y with C the MSB. The output F is logical 1 when X is greater than or equal to Y. The output F is logical 0 for all remaining values of X and Y.

Derive a minimal NOR gate only network for this system.

13. The four inputs, A, B, C, and D to a logic system represent a 4-bit binary number in which A is the MSB. If the input is less than or equal to 5_{10} the output function F is logical 1. When the input is greater than 9_{10} the output function F may be either logical 1 or logical 0. The output F is logical 0 for all remaining conditions.

Use a Karnaugh map to derive a minimal NAND gate only logic network for this system.

9 Digital storage media

9.1 Introduction

The *immediate access* memory used by a digital computer is made up of a matrix of high-speed random-access storage cells in which any particular cell, i.e., the desired storage location, can be addressed by the coordinates on the X and Y axes of the matrix. In *mainframe* computer installations, this requirement has been fulfilled by *ferrite core* magnetic storage elements. However, with the advances made in semi-conductor technology, many semiconductor memory devices have become available which are replacing ferrite cores in mainframe computers and are much more suitable for *minicomputer* and *microcomputer* memory systems.

Semiconductor memory devices include RAMs, ROMs, PROMs, EPROMs, and EAROMs and may use any of the available manufacturing techniques, e.g., bipolar, MOS, CMOS, NMOS, $I^2 L$, charge-coupled devices (CCD), magnetic bubbles, and holographic memories. However, only a limited range of devices is available using several of these techniques, and costs are relatively high. Most of the memory devices currently available use bipolar and CMOS techniques.

A microcomputer memory may be extended by using a slower bulk storage such as magnetic tape (on cassette), floppy disc, or paper tape. This is generally referred to as *backing storage*, and the devices which carry the storage media are referred to as *peripherals* to the central processing unit (CPU).

9.2 Semiconductor memory elements

A wide range of semiconductor memory elements is currently available, and certain types and packages are emerging as 'industry standards'. The main requirements of semiconductor memories are that they should occupy a small area, have a fast access time, and operate with low power dissipation. Memory devices may be either *volatile* or *non-volatile*. As the name suggests, volatile memories are changeable, i.e., when the power supply is switched off, the previously stored information is lost; and when the power supply is switched on again, there is no guarantee of the content of the memory device. Therefore, volatile memory devices are *random-access memories* (RAMs), which are frequently referred to as *read/write* (or *user*) memory elements. On the other hand, non-volatile memories are capable of retaining their stored information even when the power supply is switched off. Therefore, non-volatile memory devices are *read-only memories* (ROMs), which are widely used for storing a particular microcomputer's *operating system*, i.e., a program which the microcomputer must always obey. Further applications of ROMs include the storage of *assembler, compiler*, and *bootstrap* programs. The latter program may enable a computer to *load* programs from magnetic tape and to *dump* (or save) a computer program on to magnetic tape or floppy discs.

Bipolar manufacturing techniques offer fast access times but have a relatively low packing density, whereas unipolar techniques (MOS and CMOS) give a slower access time at much greater packing densities than bipolar. In addition, CMOS devices dissipate a very small power compared with bipolar devices. Therefore, the majority of available semiconductor memory devices use unipolar manufacturing techniques. Since memory devices are made up of regular arrays of basic cells, greater packing density is possible than for normal logic networks.

All semiconductor RAM devices may be considered as either *static*, in which the stored information is maintained as a logic state in a transistor bistable (commonly referred to as a flip-flop) arrangement as long as the power supply is switched on, or *dynamic*, in which the information is retained as a charge on a capacitor which, every few millisconds, must be subjected to a *refresh cycle* to compensate for the leakage of charge from the capacitor.

9.3 Semiconductor static RAMs

The basic cell is generally a six-transistor configuration, as shown in Fig. 9.1, in which all the transistors are *depletion mode n-channel* MOS devices. These

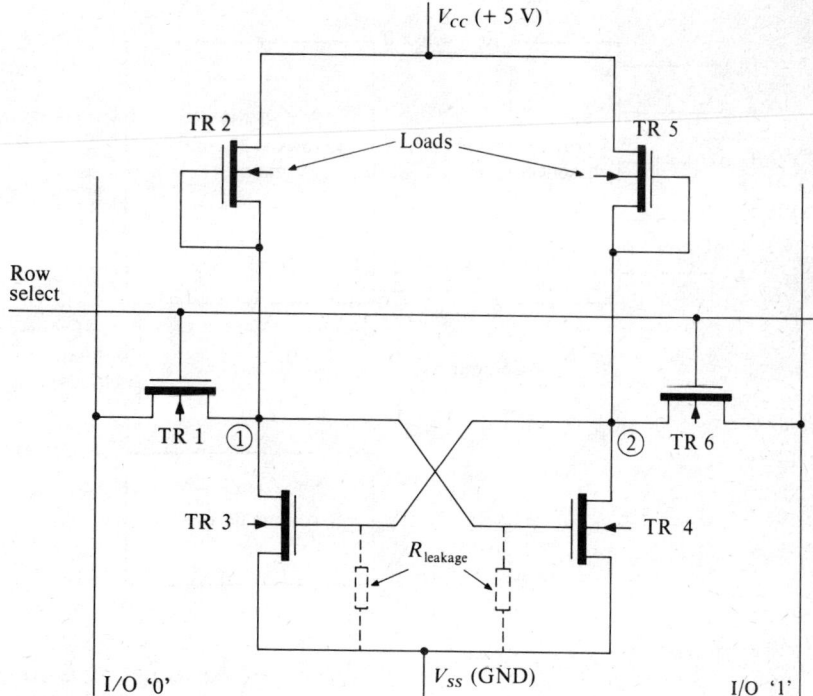

Fig. 9.1. Schematic circuit of static RAM cell. (Copyright Intel Corporation, 1978)

devices have a conduction path between source and drain even when no potential is applied to the gate, and a negative potential must be applied to the gate (relative to the source) in order to switch it off.

The basic cell operates essentially as a bistable element ('flip-flop'). Initially, assume that the potential at the gate of TR 3 is high, then TR 3 is switched on and current flows in TR 2 and TR 3. Since TR 2 is connected as a load, its impedance is greater than that of TR 3, therefore the potential at node 1 will be virtually equal to V_{SS} (GND). The same voltage is applied to the gate of TR 4, thus keeping it off. The potential at the gate of TR 3 is maintained through the load TR 5, thus holding the potential at node 2 high. This causes the load TR 5 to switch hard on, so that the potential at node 2 is virtually equal to $+V_{CC}$. The storage cell remains in this logic state (node 1 LO) until an external *write* signal is applied. The alternate logic state occurs when TR 4 is switched on, to produce a LO logic level at node 2.

The basic cells are arranged in a matrix, with data selection by the coincidence of a *row-select* line and a *column-select* line. A typical popular $1K$ static RAM chip is the Intel 2102A, which has a storage capacity of 1024 words $\times$ 1 bit housed in a 16-pin DIL package. A simplified block diagram of the 2102A is shown in Fig. 9.2, in which the storage cells are arranged in a 32×32 matrix. The *row* selection lines A_0 to A_4 access the matrix rows by means of a 1-of-32

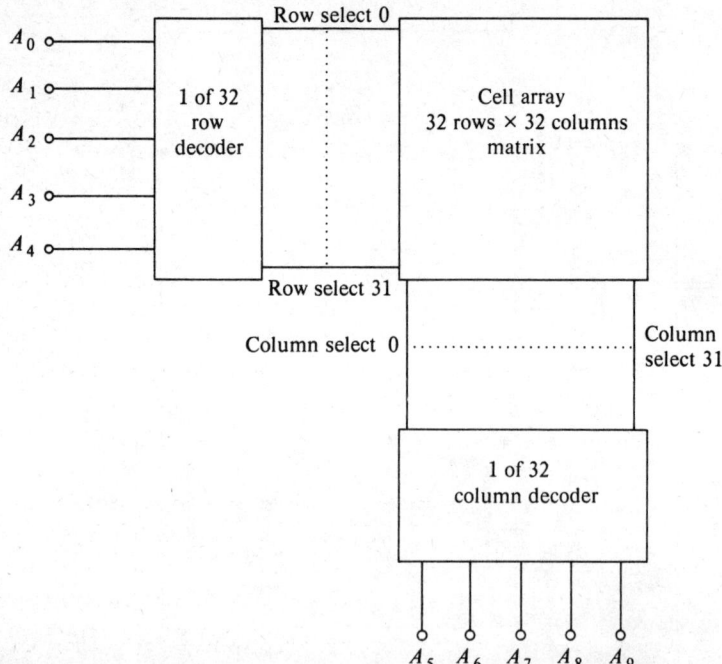

Fig. 9.2. Simplified block diagram of 2102A static RAM. (Copyright Intel Corporation, 1978).

decoder, while the *column* select lines, A_5 to A_9, access the matrix columns by means of another 1-of-32 decoder. The stored data in the selected cell is sensed, buffered, and presented to the data-out pin D_{out}. The pin configuration and logic diagram are shown in Fig. 9.3.

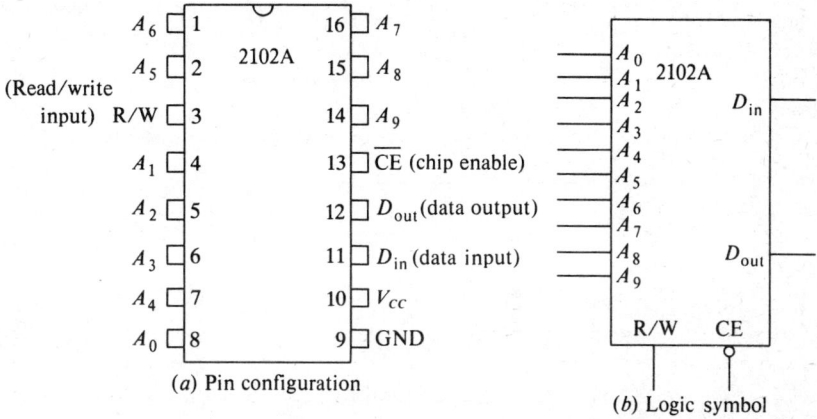

(a) Pin configuration

(b) Logic symbol

Fig. 9.3. Pin configuration and logic symbol of 2102A static RAM. (Copyright Intel Corporation, 1978).

9.4 Using the 2102A static RAM

A storage cell may be accessed for a read or write operation by applying a logical 'high' potential to the appropriate row select line which switches on transistors TR 1 and TR 6 (Fig. 9.1) of the cells in that row. For a *read* operation, a sense amplifier connected to both I/O column outputs is used to detect the state of each selected cell. A write buffer is used to apply a high level ($+V_{CC}$) on the I/O '0' line to write a logical 0, and a high level on the I/O '1' line to write a logical 1.

The 2102A static RAM incorporates internal data-in/data-out buses, and data is gated to/from the appropriate columns by column select. *'Chip-enables'* gate the output data to a tri-state buffer and then to the output pin. Therefore, if a chip is not selected, the output pin goes to a high-impedance state, thus allowing the output pins to be OR-tied. The address buffers/decoders respond to changes on the address lines and do not latch the input addresses.

Two control inputs are required: R/W (read/write) and $\overline{C/E}$ (chip-enable). For unselected inputs ($\overline{C/E}$ high), the data-in input is electrically disconnected from the internal input data bus and the data-out buffer goes to a high-impedance state. However, the addresses are buffered and decoded (generating an internal row/column select) independent of $\overline{C/E}$. The read and write cycles are as follows:

(a) *Read cycle.* The basic read-cycle timing is shown in Fig. 9.4(a). Note that although chip-enable is shown as a pulse occurring after the address changes, there is no specified time at which it must occur (either before or after address change). Therefore in cases where the data-out pin is not OR-tied

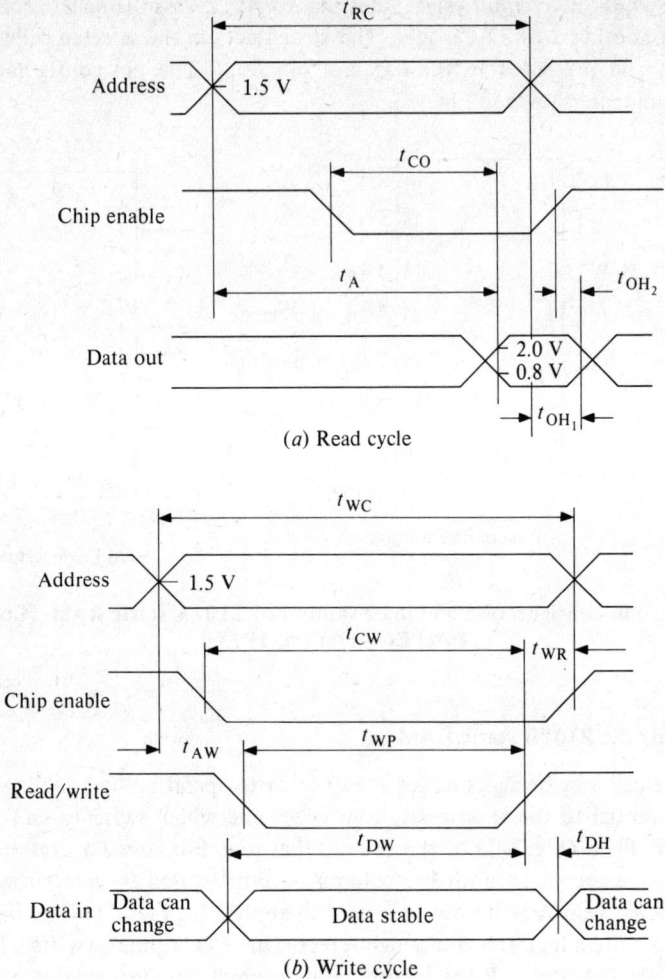

Fig. 9.4. Read/write cycle timing for static RAM. (Copyright Intel Corporation, 1978).

with other outputs, it is possible to tie the chip-enable input low and operate the RAM with only the read/write line and address inputs. Multiple read cycles may be performed by cycling through the addresses in any order. However, the read/write input must be high during this sequence, and output data will be valid at t_A (access time, typically 300 ns).

Alternatively, if the addresses are set up before a *read* decision can be made, and the chip-enable taken low at the read decision time, then output data will be valid at t_{CO} (chip-enable to output time, typically 150 ns).

(b) *Write cycle.* The basic write cycle timing is shown in Fig. 9.4(*b*). Continuous write cycles may be performed by holding chip-enable low and timing the read/write input as shown in Fig. 9.4(*b*). The minimum address-to-write set-up time t_{AW} is 20 ns, and the minimum data hold time (beyond read/write) is 0 ns. The minimum write cycle time is given by

$$t_{WC(min)} = t_{AW} + t_{WP} + t_{WR}$$

where
$$t_{WP} = \text{write pulse width (minimum 250 ns) and}$$

$$t_{WR} = \text{write recovery time (minimum 0 ns)}$$

$$\therefore\ t_{WC(min)} = 270 \text{ ns}$$

Semiconductor static RAMs are widely available in the following configurations: 256 × 1 bit, 256 × 4 bit, $1K$ × 1 bit, $1K$ × 4 bit, and $4K$ × 1 bit. Always consult manufacturer's data to obtain the connections and operating characteristics of any particular device.

9.5 Semiconductor dynamic RAMs

The basic cell in a dynamic RAM is a single transistor and a capacitor, as shown in Fig. 9.5, which includes the associated I/O circuitry. Therefore, the packing density of dynamic RAMs is about four times greater than that of static RAMs. The charge on a storage cell is gated to the bit sense line by the MOS transistor connected to the column-select line. Note that in a 64 × 64 matrix array—for a given column-select—64 storage cells are gated to the respective 64-bit sense lines.

Consider a read operation, and assume that C_{STG} is discharged, i.e., node 1 is at V_{SS} (GND). The bit sense lines are precharged to V volts (between $+V_{DD}$ and V_{SS}) by TR 1 prior to the chip-enable going high. When the address decoders have stabilized, the proper column-select line is taken high. This switches on TR 2 and connects C_{STG} to the bit sense line. The charge on $C_{I/O}$ is redistributed between $C_{I/O}$ and C_{STG}. Since $C_{I/O}$ is very much larger than C_{STG}, the change in voltage on the bit sense line is very small. The sense amplifier is designed to detect very small changes in bit sense line voltage and to latch in a state near to V_{SS} (GND) or $+V_{DD}$, depending on the state of the storage cells. During a read operation, the original charge (data) on the storage cell is changed, i.e., the read operation is a destructive read. Data is rewritten back on the storage capacitor by the sense amplifier after it has latched in the proper state.

A write operation is the same as the rewrite part of a read cycle. In this case, however, the data overrides the state of the sense amplifier and writes into the selected cell.

The basic cells are arranged in a matrix (similar to static RAMs). A typical popular $4K$ dynamic RAM chip is the Intel 2107B, which has a storage capacity of 4096 words × 1 bit housed in a 22-pin DIL package. A block diagram of the 2107B is shown in Fig. 9.6, in which the storage cells are arranged in a 64 × 64 matrix. A particular storage cell is accessed by the coincidence of a row-select (defined by addresses A_0 to A_5) and a column-select (defined by addresses A_6 to A_{11}). An on-chip timing and control generator provides for the internal timing signals for decoding, read/write strobing, data, and output gating. All timing circuits are activated by a positive-going edge of chip-enable.

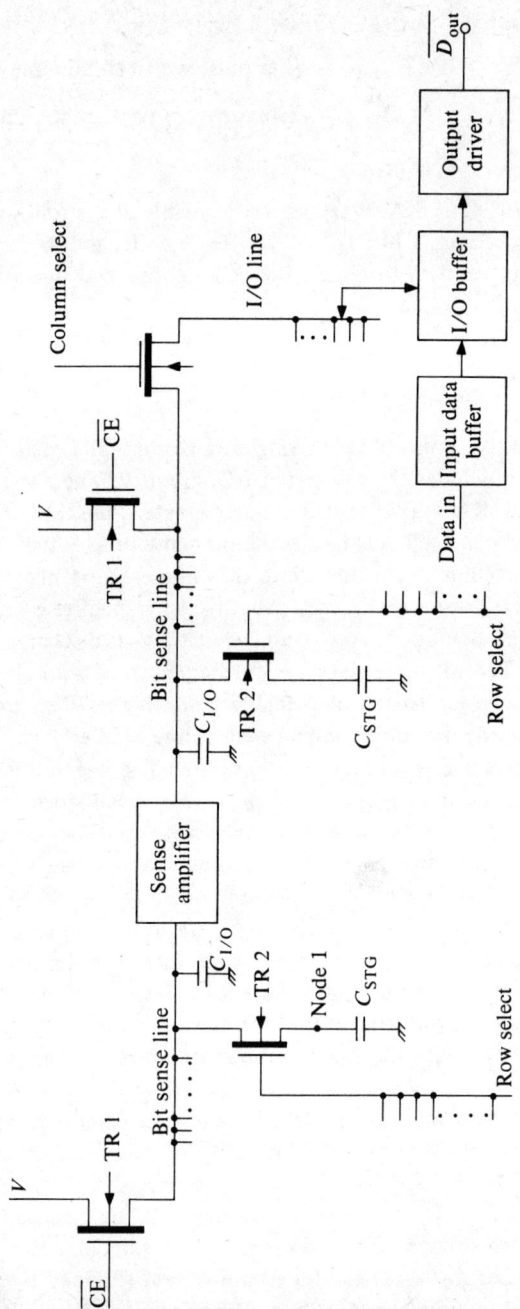

Fig. 9.5. Schematic circuit of basic dynamic RAM cell and associated I/O circuitry. (Copyright Intel Corporation, 1978).

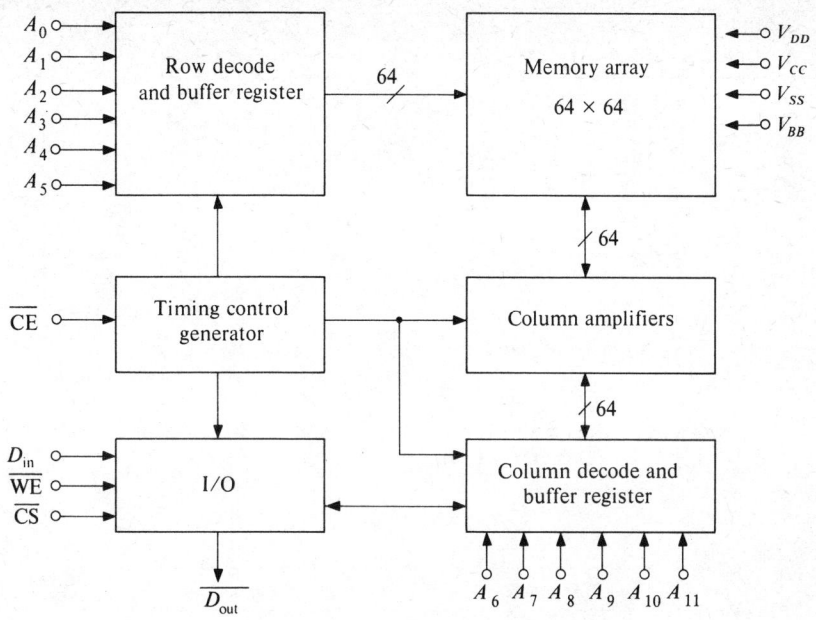

Fig. 9.6. Block diagram of Intel 2107B dynamic RAM. (Copyright Intel
Corporation, 1978).

Chip-select controls the internal data I/O gating circuits in the 2107B. When
chip-select is high, the output data buffer is in a high-impedance state and the
data-in buffer is electrically isolated from the data-in input pin. Since chip-select
controls only the internal data buffers and not the internal timing generators or
address buffers, it is possible to refresh the 2107B with chip-select high by
initiating a read/refresh or write cycle.

The address buffer registers consist of latches activated at the leading edge
of chip-enable. Since the addresses are latched shortly after chip-enable goes high,
it is permissible to change the address long before the memory cycle is completed to set up for the next cycle.

The write-enable input activates the data-in buffer-gating data to the selected
memory cell. Input data must be valid at the time write-enable goes low to
ensure that the proper data is written into the memory. The pin configuration
and logic symbol are shown in Fig. 9.7.

9.6 Using the 2107B dynamic RAM

The timing relationships for the read/refresh cycle are shown in Fig. 9.8. It is
imperative, for all cycles, to ensure that the address inputs are valid at or before
chip-enable reaches the $(V_{SS} + 2.0)$–volt level (t_{AC}). The high speed of the
internal address buffer/latches means that if the address inputs are not valid until
just after the chip-enable goes high, the incorrect address is likely to be latched

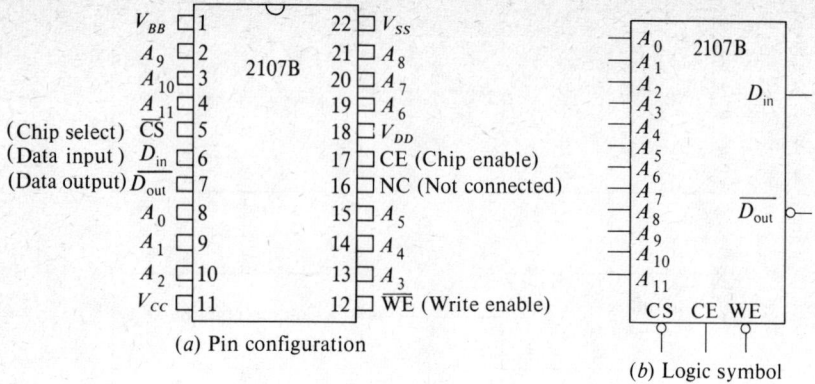

(a) Pin configuration

(b) Logic symbol

**Fig. 9.7. Pin configuration and logic symbol for 2107B dynamic RAM.
(Copyright Intel Corporation, 1978).**

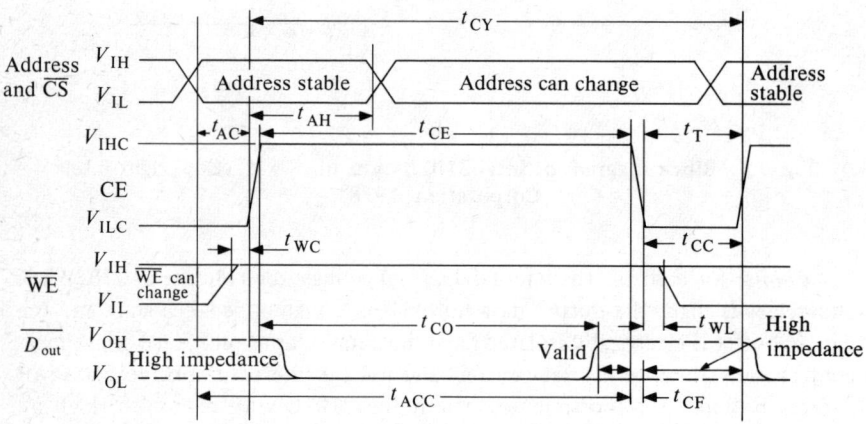

**Fig. 9.8. Read/refresh cycle timing for 2107B dynamic RAM.
(Copyright Intel Corporation, 1978).**

in the chip. Similarly, the input data must not change after write-enable goes low while chip-enable is high. For all cycles, the data-out output goes low shortly after chip-enable goes high. This prevents the output from being tied directly to a preset or clear input of a latch.

The 2107B dynamic RAM chip is not completely TTL-compatible. For example, the high level of the chip-enable must be at least 11 V. The required levels may be obtained by using a special driver chip. A suitable chip is the Intel 3245, which is a quad MOS *level-driver*, as shown in Fig. 9.9. Each driver is capable of driving a 250 pF load with a maximum delay of 30 ns.

The most serious problem encountered when using dynamic RAMs is due to transients. However, in the 2107B, these transients may be easily handled provided proper attention is paid to peak values and adequate decoupling is used.

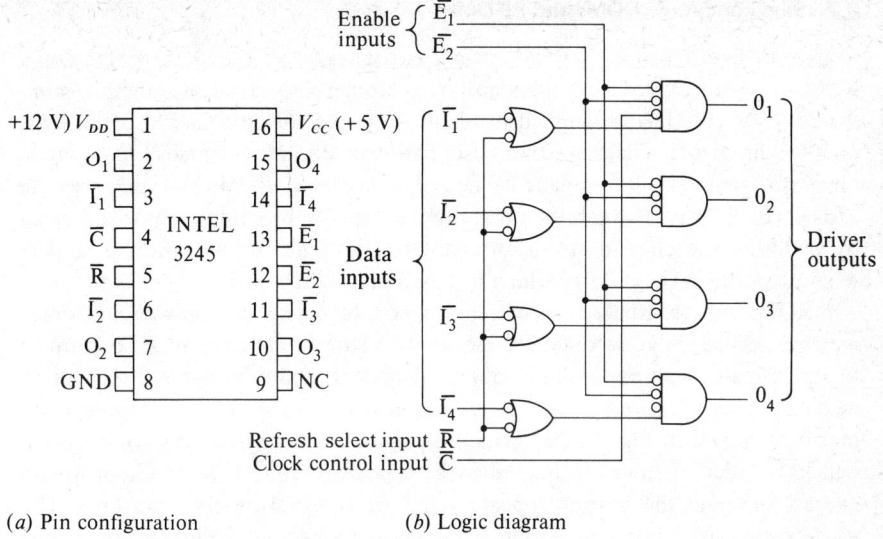

(a) Pin configuration   (b) Logic diagram

Fig. 9.9. The Intel 3245 quad MOS level driver. (Copyright Intel Corporation, 1978).

Semiconductor dynamic RAMs are widely available in the following configurations: $4K \times 1$ bit, $8K \times 1$ bit, $16K \times 1$ bit, and $64K \times 1$ bit. Always consult manufacturer's data to obtain the connections and operating characteristics of any particular device and its associated driver and refresh requirements.

Charge-coupled-device (CCD) technology has provided another type of dynamic RAM. These devices have been referred to as 'bucket brigade' devices, in which the 'buckets' represent the storage cells which, depending on the stored information, either contain a charge or not. In CCDs the storage cells are arranged in long looped lines so that the charges are passed from cell to cell. Read, write, and refresh operations are performed as the relevant information passes that particular storage location. These devices therefore store information in serial form, i.e., one bit after another, and require much less associated circuitry for their operation, since the need to address each cell separately is eliminated. CCD memories can store large amounts of information—typically, $64K \times 1$ bit in a 16-bit DIL package—although access time may be relatively long.

One method of increasing the speed is to arrange for several loops, e.g., a $64K \times 1$ bit CCD memory chip housed in a 16-pin DIL package may be arranged in sixteen addressable $4K$ blocks, i.e., requiring only four address lines. Access time is typically 250 ns.

Magnetic bubble memories circulate magnetized areas in a similar way to CCD memories, but at present very few of these chips are available.

9.7   Semiconductor ROMs and PROMs

Several different forms of ROM are currently available: the *mask programmed* ROM is manufactured to a standard or customer's specification, the *programmable* ROM (PROM) is supplied in blank form and the customer 'programs' his own requirements. There are two basic forms of PROM: the *fusible link* which, once programmed, is permanent, i.e., it becomes a ROM; and the *erasable* PROM (EPROM), in which the programmed data can be completely wiped clean by exposing the chip to strong ultraviolet (UV) light, usually through a glass window in the package, after which it may be reprogrammed.

ROMs are essentially low-cost, high-speed, high-density, non-volatile storage matrices. Since they are generally required to store a pre-determined pattern of logical signals, such as 'look-up tables', decoders, and converters, they can be made up of combinational logic networks instead of 'flip-flops'. A simple diode matrix is shown in Fig. 9.10(a), which is part of a denary-to-binary converter, in which the denary number is applied as a positive voltage to the appropriate address line and the output appears as a logical signal on the data lines. The diode pattern and interconnection are achieved by varying the masks used in the diffusion stages during manufacture. A transistor array using MOS techniques (which corresponds to the diode array) is shown in Fig. 9.10(b). In this case, only *selected* transistor gates are connected to the address lines, so that a logical 1 (i.e., negative voltage) signal input to an address line causes the transistors with their gates connected to be turned on, thus providing a logical 0 on its output data line.

Fusible-link PROMs are available in two forms. In the first, a matrix of diodes (or transistors) is formed with a fusible link in series with each diode. Programming is achieved by fusing the links (by passing a controlled current pulse through the link) in series with the diodes which are *not* required in the matrix, as shown in Fig. 9.11(a). Once the required bit pattern is programmed (written into memory) it cannot be changed. Material used for the fusible links includes polycrystalline silicon, platinum silicide, nichrome, and titanium tungsten. The second form of fusible-link PROM is made up of a matrix as shown in Fig. 9.11(b), in which each cell is formed by two diodes connected back to back, thus making open-circuit cells initially. Programming is achieved in this case by applying a sufficiently high voltage and passing a current pulse through the cells that *are* required in the matrix to break down the reverse-biased diode.

Ultraviolet erasable PROMs (EPROMs) are sometimes referred to as *read-mostly memories* since they can be reprogrammed following erasure by UV radiation. Each cell comprises an MOS transistor having two gate electrodes, as shown in Fig. 9.12(a). The *control* gate is connected to row-decoding circuitry and the lower gate 'floats' in an insulating sea of silicon dioxide. The turn-on threshold voltage at the control gate is controlled by the charge in the floating gate. With no charge in its floating gate, the selected transistor conducts. When the floating gate is charged, the selected transistor does not conduct. Programming is effected by connecting the source and substrate to ground, and applying a voltage (typically 25 V for 50 ms) to the control gate and drain. This voltage

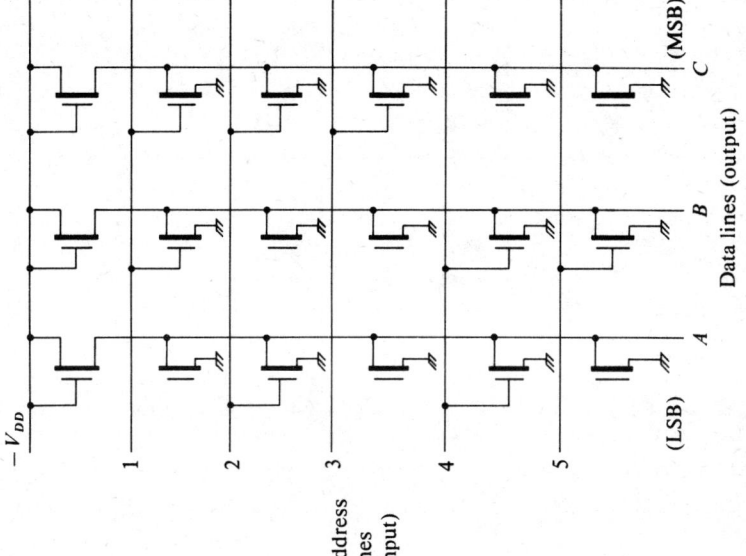

(b) MOS transistor matrix

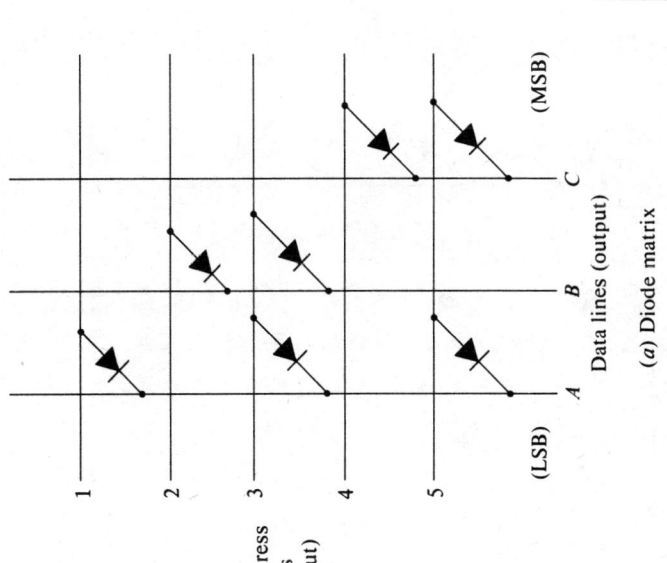

(a) Diode matrix

Fig. 9.10. Mask-programmed ROMs (denary-to-binary converter).

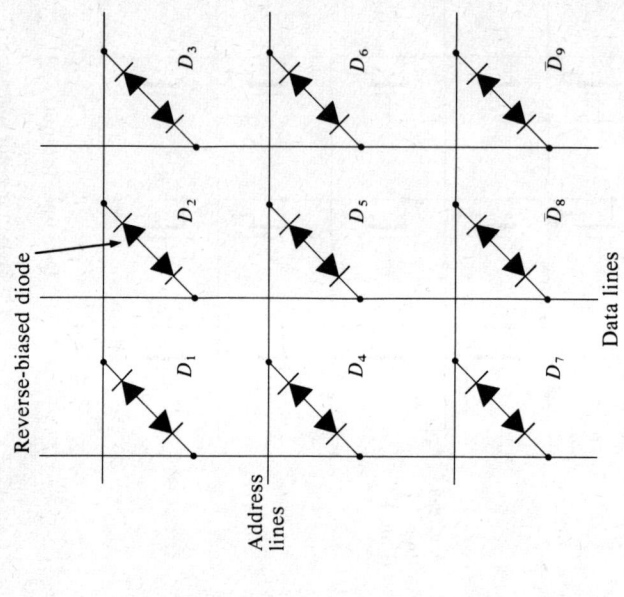

(b) Reverse-biased diodes

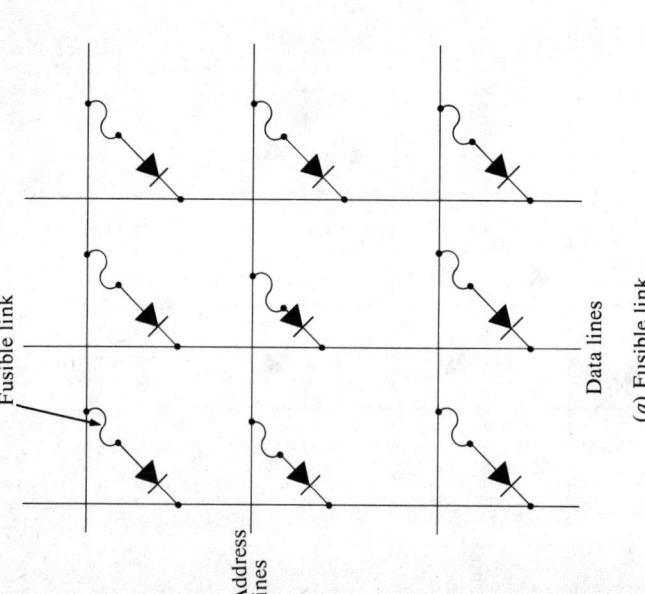

(a) Fusible link

Fig. 9.11. Fusible-link PROMs.

causes some of the electrons in the silicon dioxide insulation to become trapped in the floating gate, where they remain indefinitely. Ultraviolet radiation increases the conductivity of the silicon dioxide which allows the trapped floating gate charge to leak away from all such charged cells in the chip.

The *electrically-alterable* PROM (EAROM) is similar to the EPROM, but uses a metal nitride–oxide semiconductor (MNOS) device as shown in Fig. 9.12(*b*).

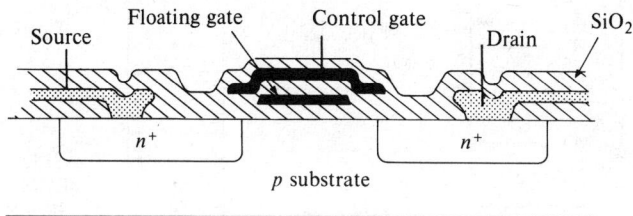

(*a*) Ultraviolet erasable MOS cell

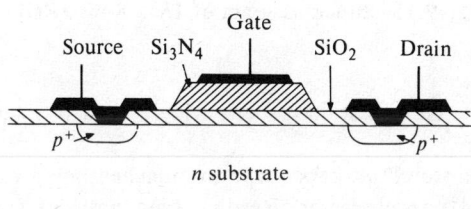

(*b*) Electrically alterable MNOS cell

Fig. 9.12. Erasable PROMs.

The interface between the silicon nitride and silicon dioxide behaves in a similar way to the floating gate. This device is programmed by applying a large field to the control gate, which causes the dielectric interface to become charged. A reverse field application causes the interface to become discharged. Therefore, the EAROM can have individual cells erased and reprogrammed without affecting the rest of the chip.

EPROMs are now more widely used than mask-programmable PROMs. This is largely due to their reducing cost, increased availability, and convenience of use. The basic block diagram of an $8K$ PROM is shown in Fig. 9.13, which is organized as $1K \times 8$ bits.

9.8 Paper-tape storage

Paper tape, about 1 inch wide, provides a bulk storage method which is used in some microcomputer systems. Many teletypewriters (TTYs) incorporate a paper-

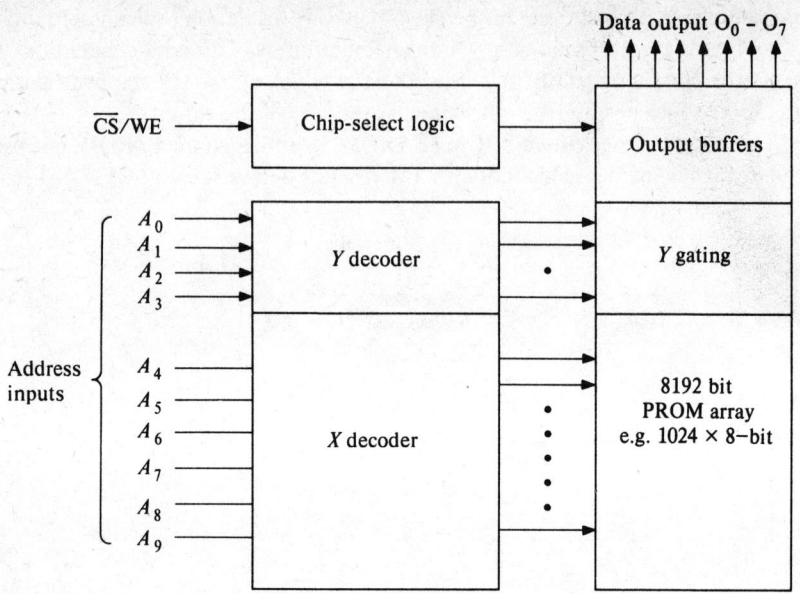

Fig. 9.13. Block diagram of $1K \times$ 8-bit PROM.

tape reader/punch, and these have been widely used in mainframe computer installations.

Information is stored on paper tape by punched holes, each row across the paper representing one character. Reading (and writing) from paper tape is relatively slow at 10 to 100 characters per second, since this is a *serial* memory. It is therefore difficult to change a character, or group of characters. The most widely used code on paper tape is ASCII (*A*merican *S*tandard *C*ode for *I*nformation *I*nterchange), as shown in Fig. 9.14.

9.9 Magnetic-tape storage on cassette

The domestic cassette tape recorder provides an inexpensive bulk storage method but, due to the slow data transfer, i.e., *serial* access, is generally restricted to hobbyist microcomputer systems. Equipment is necessary to convert the digital signals to audio tones and vice versa—this is a *modulator/demodulator* (MODEM). MODEMs use the principle of *frequency-shift keying* (FSK), in which a burst of tone at one frequency represents a logical 1, and a burst of tone at another frequency represents a logical 0. The most widely used standard for magnetic tape is the *Computer User's Tape System* (CUTS), which may also be referred to as the *Kansas City* or *Byte Format*. This system records data at 300 baud, with a logical 1 recorded as eight cycles of a 2400 Hz tone, and a logical 0 recorded as four cycles of a 1200 Hz tone. A byte of data is then recorded as a start bit of logical 0, followed by eight data bits, then two bits of logical 1.

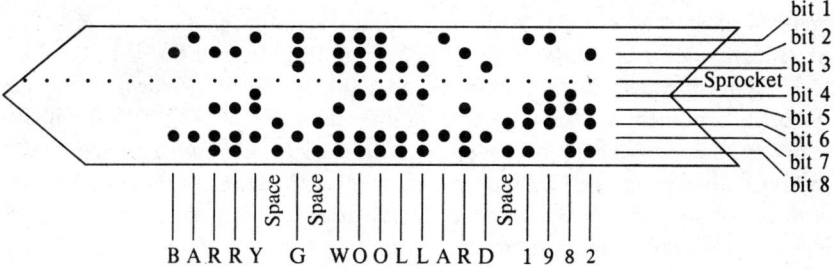

(*a*) Coded paper tape

Char.ᶜ	Code 87654321	Char.ᶜ	87654321	Char.ᶜ	Code 87654321
A	01000001	Y	01011001	'	10100111
B	01000010	Z	01011010	(	00101000
C	11000011	0	00110000	)	10101001
D	01000100	1	10110001	*	10101010
E	11000101	2	10110010	=	10111101
F	11000110	3	00110011	←	11000000
G	01000111	4	10110100	[	11011011
H	01001000	5	00110101	\	01011100
I	11001001	6	00110110	+	00101011
J	11001010	7	10110111	↑	11011110
K	01001011	8	11011000	]	10111000
L	11001100	9	00111001	<	00111100
M	01001101	:	00111010	>	10111110
N	01001110	−	00101101	?	00111111
O	11001111	;	10111011	TAB	01011111
P	01010000	,	10101100	SPACE	10100000
Q	11010001	.	00101110	LINE FEED	00001010
R	11010010	/	10101111	RETURN	10001101
S	01010011	!	00100001	ESCAPE	00011011
T	11010100	"	00100010	RUBOUT	11111111
U	01010101	#	10100011	BREAK	11111100
V	01010110	$	00100100		
W	11010111	%	10100101		
X	11011000	&	10100110		

EVEN PARITY.

(*b*) Code table (ASCII 8-bit)

Fig. 9.14. ASCII 8-bit code.

9.10 Floppy-disc storage

The floppy-disc system provides the ultimate in bulk storage techniques, the name being derived from the fact that the discs used are made of a flexible material, coated on one side with a smooth surface of magnetic oxide. Two sizes of disc are widely available—$5\frac{1}{4}$ in (13.335 cm) diameter and the diskette of 8 in (20.32 cm) diameter. Data is recorded on a floppy disc in the form of magnetic pulses in concentric *tracks*, e.g., an 8 in diskette may have 77 tracks (numbered 0 to 76). Some discs may be coded on both sides (*double-sided*). Each track is divided into *sectors*, e.g., an 8 in disc may have 26 sectors (numbered 1 to 26), each sector storing 128 characters of information. Some systems may store 256 characters in each sector—these are referred to as *double-density* discs. One or more index holes are punched in the disc to enable the controller and drive system to detect the start of the sectors. *Soft-sectored* discs each have only one punched hole to identify the start of the tracks. The start of each *sector* is determined by a calculation of the time interval since the index hole passed the read/write head, as shown in Fig. 9.15(*a*). *Hard-sectored* discs have index holes punched between adjacent sectors.

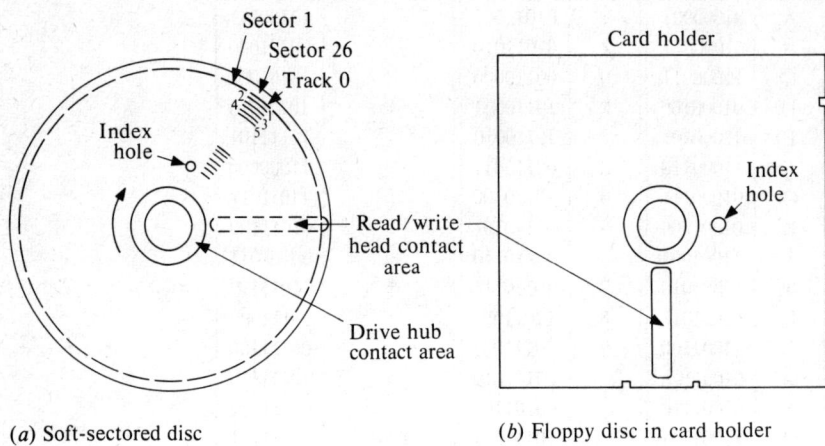

(*a*) Soft-sectored disc   (*b*) Floppy disc in card holder

Fig. 9.15. Floppy disc.

The disc is supported in a card envelope with a slot cut out to allow the read/write head to record data on the magnetic surface, as shown in Fig. 9.15(*b*). The disc is driven from the centre at approximately 360 r/min, and the read/write head moves in and out over the surface in increments of one track.

A single-sided, single-density 8 in disc can therefore store 77 (tracks) × 26 (sectors) × 128 (characters per second) = 256 256 characters, which is approximately 31*K* bytes, and this random-access storage has an access time of approximately 200 ms.

Floppy discs must be *formatted* before they can be used. Formatting is the process by which the microcomputer is used to code the disc to identify the tracks and sectors. A series of identification marks is coded at the beginning of each sector so that the floppy-disc controller can identify its location. These codes are compared with the required sector address to determine if there is an error. Codes are also included in the *'preamble'* to test the read/write head operation. Each sector then has 128 character positions, followed by futher testing codes (*'postamble'*). When the disc has been formatted, instructions and programs may be written to and read from the disc.

9.11 Ferrite core store

The ferrite core store basically consists of matrices of ferrite cores threaded on wires. The cores are grouped to form *locations*, each location being a *computer word* of a suitable size to store items of data or program instructions. The size of the word used depends upon the computer, but is generally in the range 8 to 48 bits (cores). The total storage capacity is generally quoted as a number of K words, where $1K$ represents 1024 words (i.e., 2^{10}). Each core is of approximately 0.4 mm diameter (0.015 in).

The ferrite core can be magnetized in either direction in terms of *two* remanent conditions of magnetization, by passing a steady current of I_m amperes through one wire and then removing it, as shown in Fig. 9.16 (i.e., either positively or negatively). Each core in the matrix has an X wire and a Y wire passing through it, so that it is possible to *select* any particular core by specifying the appropriate X and Y wires and passing *half* the current ($\frac{1}{2}I_m$) necessary to magnetize a core, as shown in Fig. 9.17.

The matrices of cores are generally arranged in planes such that the corresponding core in each plane is used to represent a computer word. An additional plane is used for a *parity* check, as shown in Fig. 9.18, for an 8-bit word (8-bits = 1 byte).

Each core in the matrix has *four* wires passing through it, which are identified as follows:

1. X wire.
2. Y wire.
3. Z wire (inhibit).
4. Sense (read) wire.

In each matrix there is a *single* Z wire and a *single* read wire, both of which pass through every core in that matrix as shown in Fig. 9.19.

9.12 Reading and writing in store

When a current equal to half the value required to magnetize a core is applied simultaneously to the selected X and Y wires, the resulting effect is a current of sufficient value to magnetize the selected core at the intersection of the two wires. The four operations which are performed in a core store are as follows:

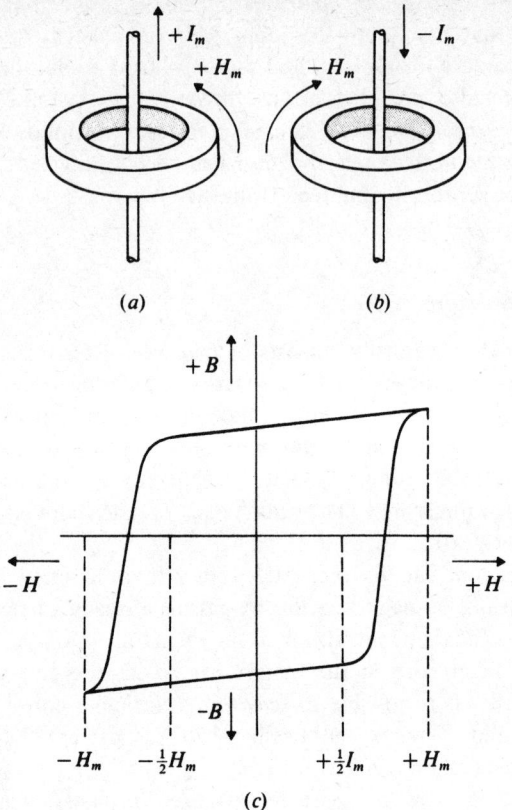

Fig. 9.16. Ferrite core magnetization. (*a*) Positive magnetization, (*b*) negative magnetization, (*c*) magnetic characteristic.

(a) *Read cycle.* The state of a core is read by *writing* a logical 0 state to that core and detecting the change, if any, in its state. The read wire S, threaded through each core in the matrix, is used to detect the change. If the state of the core changes, a current is induced in the S wire, i.e., if the original state was 0, there is no current, while if the original state was 1, a current is induced.

(b) *Regenerative cycle.* After the read cycle all the selected cores will be in the 0 state, so that if the information originally stored needs to be retained, it is necessary to regenerate that information in the cores.

The selected X and Y wires have passed through them a reverse current of sufficient value to magnetize the appropriate cores to the 1 state. However, some of these cores may *originally* have been at logical 0. To prevent these cores being changed to 1 the Z (*inhibit*) wire is used, i.e., when a 0 is read a current is fed to the Z wire coincidentally with the half-currents in the X

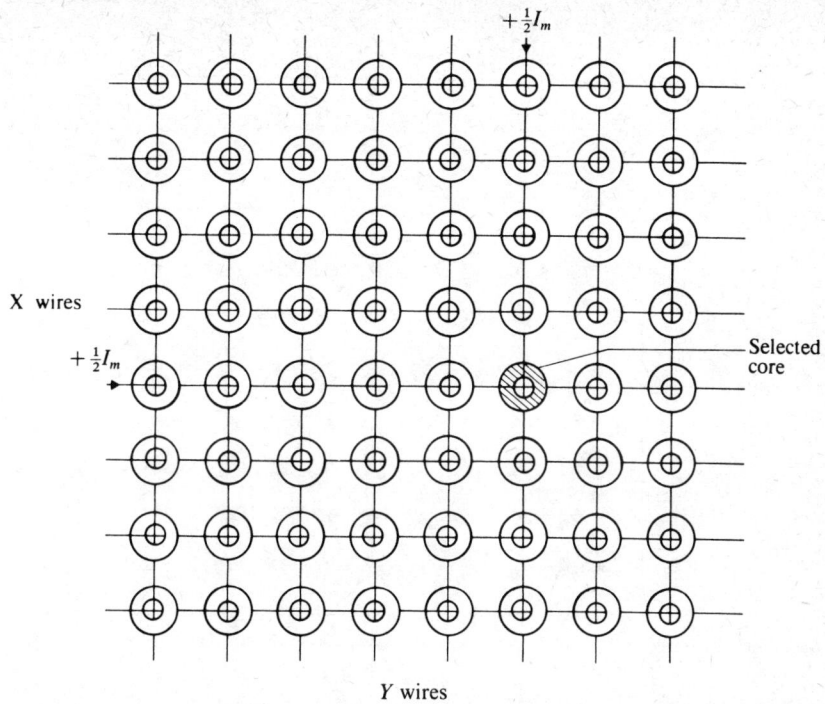

X wires

$+\frac{1}{2}I_m$

$+\frac{1}{2}I_m$

Selected core

Y wires

Fig. 9.17. Selection of a core in a matrix.

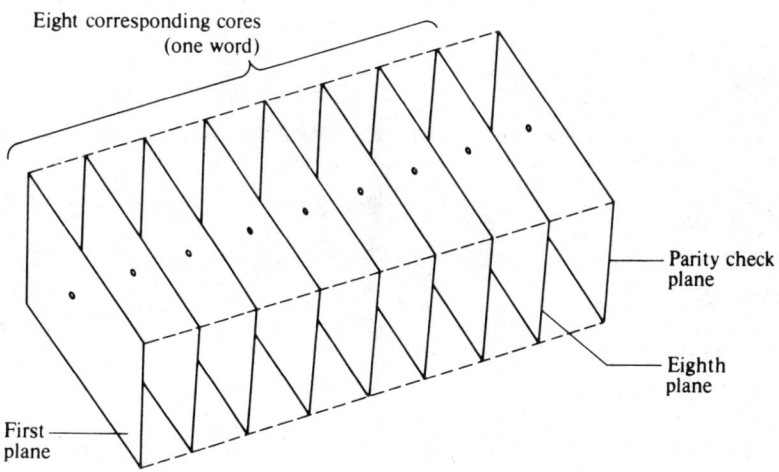

Eight corresponding cores
(one word)

First
plane

Parity check
plane

Eighth
plane

Fig. 9.18. Representation of an 8-bit computer word.

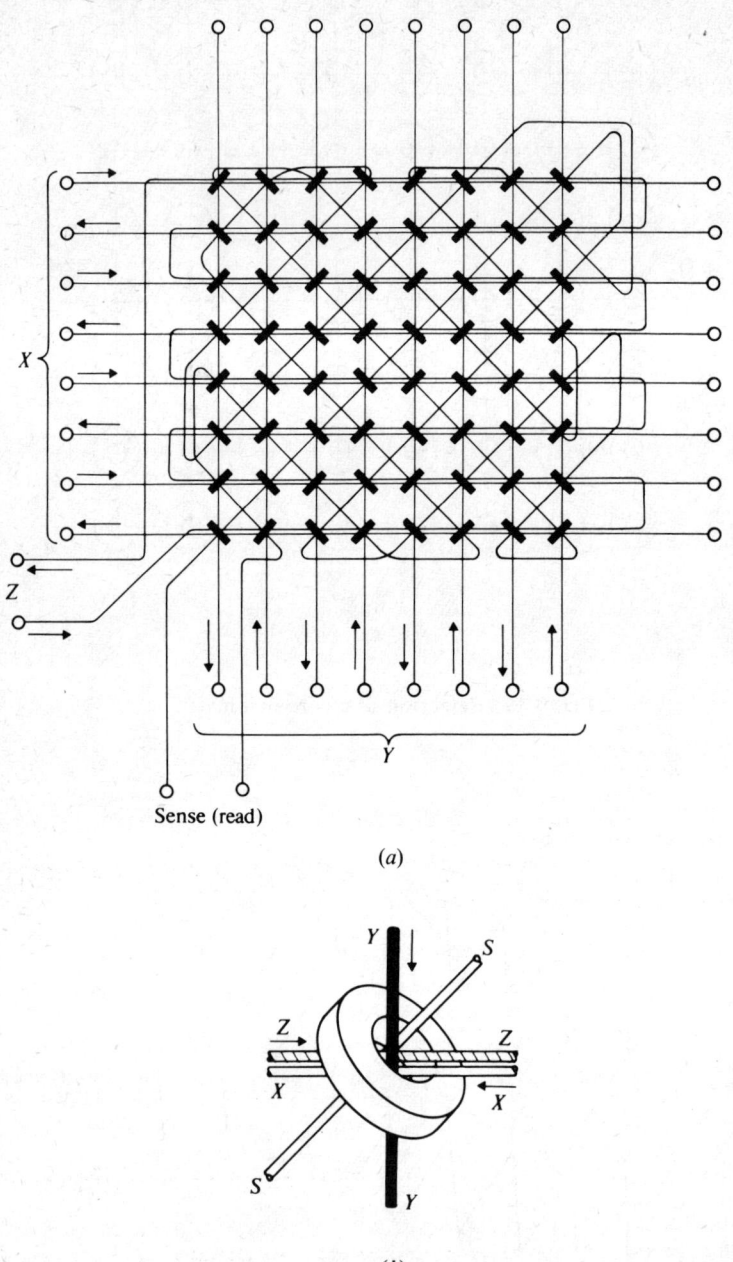

Fig. 9.19. Ferrite core matrix array with wires. (*a*) Matrix, (*b*) individual core.

and Y wires so that the resulting current is insufficient to change the core state from 0 to 1.

(c) *Write cycle.* This is a similar process to regeneration, with the Z wire current being derived from the information to be written. The write cycle is always preceded by the read cycle.

(d) *Parity check.* This is a method by which any transfer of data can be checked. *Odd-parity* is used to indicate that the total number of 1s in the word (including the parity bit) is always *odd*. *Even-parity* is used to indicate that the total number of 1s in the word (including the parity bit) is always *even*.

During each cycle the parity of the word is checked to ensure that odd-parity (most common) is maintained, thus indicating a successful transfer.

Core store is therefore a direct-access system—each location has a numbered *address*. The time taken to access information (*access time*) is better than 2 μs and is the same for any location. This type of memory has high reliability, low power consumption, and occupies a relatively small space.

EXERCISES TO CHAPTER 9

1. Explain the meaning of the terms *volatile* and *non-volatile* in relation to digital storage media.

2. Briefly explain the meaning of the following terms:
   (a) random access memory (RAM);
   (b) read-only memory (ROM).

3. Briefly explain, with the aid of sketches, how a required storage location is *addressed* in any storage medium.

4. Describe how data may be written into a selected store location in a RAM.

5. Describe one method which can be used to read data from a selected storage location in a memory system.

6. Explain the meaning of the following terms:
   (a) programmable read-only memory (PROM);
   (b) erasable programmable read-only memory (EPROM);
   (c) electrically alterable read-only memory (EAROM).

7. Sketch a block diagram of a typical semiconductor RAM, and briefly explain how it performs its function.

8. Sketch a block diagram of a typical semiconductor ROM, PROM, or EPROM, and briefly explain how it performs its function.

9. Explain the meaning of the terms *static* and *dynamic* as applied to semiconductor RAMs.

10. Explain, with the aid of sketches, how ferrite cores may be used to store digital data.

11. Sketch a single-plane 4 × 4 ferrite core matrix and use it to describe coincident current addressing of a particular location.

12. Describe the operations *read cycle, regenerative cyle*, and *write cycle* in a ferrite core matrix storage system.

10 Circuit techniques and fault diagnosis

10.1 Introduction

The growth rate of the development and application of integrated circuits (ICs) during the last 25 years has had a profound effect on the design and construction of electronic circuits and systems. Early circuit solutions required complex wiring harnesses which contributed as much as half of the total volume in some applications. The techniques of production of printed circuit boards (p.c.b.s) were rapidly developed to accommodate the increasing degree of complexity of modern circuits until, by about 1960, the production of multilayer p.c.b.s became possible—a technique which encouraged increased packing densities and, at the same time, proved to give highly reliable interconnections. The use of printed circuits can be extended from the connection of circuit elements to the interconnection of circuit modules by mounting a connector (e.g., p.c.b. edge connector) to a *mother-board* and plugging-in *daughter-boards* at right-angles to it. Modular equipment design is increasing in popularity, probably due to its suitability for production flow-line assembly of easily tested and inspected modules. This technique also lends itself more readily to modification or variation.

The quality of manufactured components is determined by the design, materials and manufacturing processes of the components. Reliability is one of the means by which the quality of a component, or of equipment, is measured. The two main factors to achieve a high degree of reliability in semiconductor-based devices are design and the control of the quality of materials and of the manufacturing processes. Testing is, of course, an important element in checking that each specimen at the completion of each stage of manufacture is what it is supposed to be. The reliability of discrete and IC semiconductors is very dependent upon technological progress in their design, in their method of manufacture and in the methods of their use. Equipment designers unfortunately rarely have time for the detailed considerations which reliability quantities demand, so that indiscriminate failure rate figures have been applied to groups of devices.

It is extremely important for the technician to develop the ability to diagnose the causes of faults in electronic circuits and systems. This expertise is not acquired easily, but is built on a foundation of understanding of components and circuit operation together with knowledge of test and measuring equipment and techniques of testing methods.

10.2 Breadboarding

Solderless breadboards provide a convenient flexible method for prototype circuit development using ICs and/or discrete components. The solderless breadboard is generally made up of a plastic block housing, say 47 rows of five interconnected contacts arranged in a 0.1 in matrix, as shown in Fig. 10.1(a). The standard layout of a typical breadboard is shown in Fig. 10.1(b), in which rows of continuous contacts are incorporated at the top and bottom which are ideal for use as power supply rails.

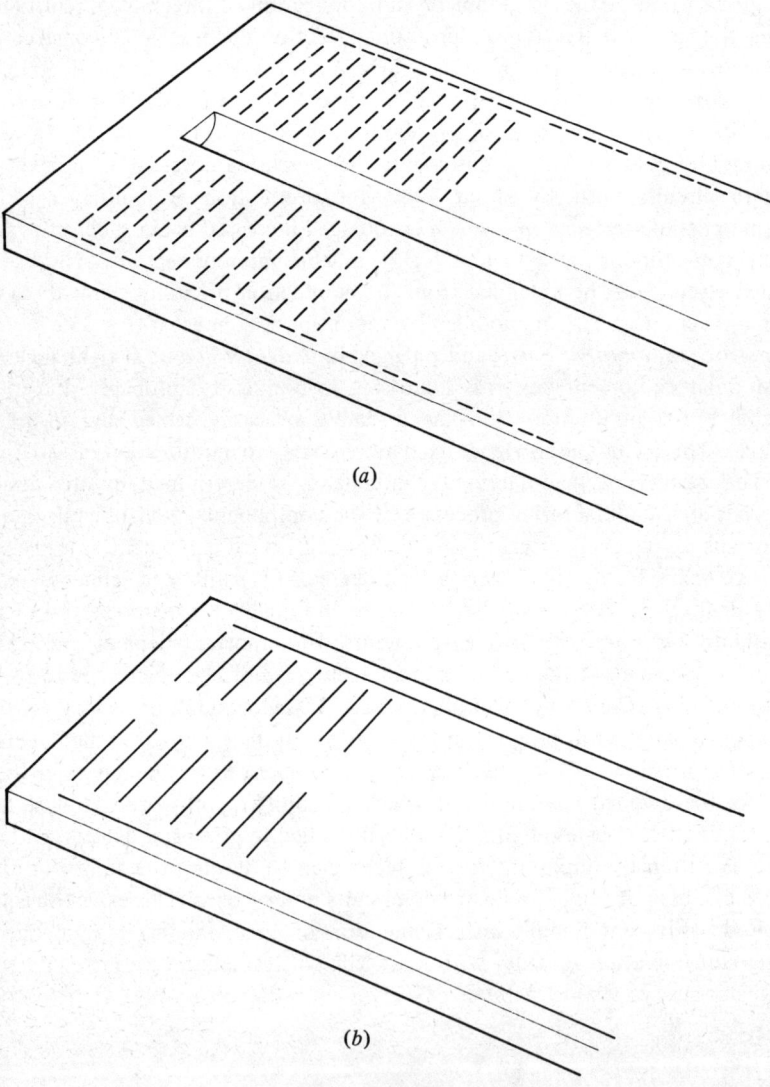

(a)

(b)

Fig. 10.1. Solderless breadboard.

It is strongly recommended that a neat and orderly layout is adopted when using breadboards. Decoupling capacitors—typically 0.1 μF—must be used, at least one capacitor for every two or three ICs, connected between $+V_{CC}$ and GND. A rectangular layout is recommended for wire interconnections—it may take a little longer, but is much easier to check and test. Wire interconnections are best made by using single-strand (typically 0.6 mm dia) wire with coloured PVC insulation, so that circuit wiring may be colour-coded for ease of identification. A selection of prepared cut lengths of wire reduces circuit construction time and helps to maintain continuity of construction.

A useful technique is to keep accurate records of circuits which have been breadboarded: note the results of all measurements made, and carefully record any changes or variations made to the circuit and their effects upon circuit operation.

The *Digital Trainer*, shown in Fig. 10.2, is manufactured by Beal-Davis Electronics Ltd, and is equipped with a large breadboard area and generous d.c. power supplies as follows: 5 V, 5 A; 12 V, 1 A; ±15 V, 1 A each rail; thus enabling digital, linear and interface circuits to be developed and tested. Binary signal 'toggle' switches, with 'de-bounced' outputs, are incorporated to provide

Fig. 10.2. The Digital Trainer. (with permission of Beal-Davis Electronics Ltd.)

the logical signals '1' and '0', with two 'initiation' push-button signal switches to give a logical 'LOW to HIGH' transition and a logical 'HIGH to LOW' transition when pressed. A variable, low-frequency (approx. 1 Hz to 20 Hz) clock pulse generator with an LED indicator at its output is available to carry out tests on logic circuits at a slow speed enabling the observer to follow the transitions of logic states through the circuit. A fixed-frequency pulse generator (approx. 2 kHz) is available as a faster clock pulse generator. Logic state indicators in the form of LEDs are fitted to facilitate monitoring of logic states during development and testing of digital circuits. Finally, seven-segment LED displays are incorporated—both common-anode and common-cathode—thus enabling digital counters to be more easily developed and tested.

10.3 Stripboard

Copper strips bonded to SRBP board and drilled, at say 0.1 in centres, to accept terminal pins and component leads. This provides a simple form of printed circuit, with the copper strips forming the interconnections which may be cut where breaks are required by using a stripboard cutter.

10.4 Printed circuit board (p.c.b.)

Printed circuits generally provide an economical method for mounting and inter-connecting ICs and discrete components. Standard prototype boards are widely available which are developments of the stripboard, in which copper track is bonded to SRBP board or epoxy glass material conforming to a range of require-ments, e.g. a general layout suitable for a 'mix' of discrete components and DIL ICs and may also be equipped with 'pads' for edge connection. Other standard circuit boards have tracks and pads bonded on the board to layouts which con-form to widely used microprocessor bus systems, e.g., S 100.

P.c.b.s are produced using several different techniques, the most popular for small-scale production being the *direct method* and the *photographic method.* The former uses a copper-clad epoxy glass board on which the *artwork* is applied directly, either by using a pen filled with etch-resist ink or by using the wide range of available etch-resist p.c.b. transfers (including pads, tracks, connectors, right-angled D connectors, DIL pads, transistor pads, links, etc.). The board, with its artwork of the p.c.b. layout required attached, is then etched in a solution of ferric chloride. After etching, the resist ink and/or transfers are removed, and the board may be drilled ready for component mounting. With the photographic method, special photo-resist copper-clad board is used. The artwork is prepared by applying p.c.b. resist transfers and/or resist ink onto a sheet of translucent polyester film—alignment of the layout may be aided by using a transparent 0.1 inch layout grid. This artwork is then placed against the positive photo-resist surface of the board and exposed to UV light for several minutes, then removed for developing in a solution of sodium hydroxide. Finally, the board is etched in a solution of ferric chloride. The artwork generally used for this small-scale production is on a 1:1 scale. As the degree of complexity of p.c.b.

layout increases, it is useful to use larger-scale transfers such as 2:1 or even 4:1 to prepare the artwork. The enlarged artwork is then photographically reduced to standard size for transferring onto the board. This technique achieves a greater accuracy and more easily defined track spacing than is possible using a 1:1 scale.

Designers of printed boards for ICs should always aim at obtaining the best possible packing density. Poorly designed, poorly laid out boards cost just as much to produce as well-designed boards, but well-designed boards will minimize interconnection lengths and will generally improve the board's design function. Packing density will be minimized in those applications where the power dissipation is relatively high in order to allow natural air circulation to dissipate the heat generated by the ICs. The designer must also consider such factors as ease of fault finding, since the time taken to locate and replace a defective IC may well affect the value and the number of spares to be held, and also the way in which the system is broken down into its sub-assemblies.

Multi-layer p.c.b.s invariably use an epoxy resin glass fibre as the base material, and are made up from a stack of single-sided etched sheets, laminated by heat and pressure to produce a single board. Experience has shown that the best packing densities and the simplest board layouts are achieved using *X–Y* co-ordinate wiring, where all the tracks on one side of the board generally run in one direction, and the tracks on the other side of the board run at right angles to them.

Various methods are used for the interconnection of the layers of multi-layer boards, one of the most popular being the *plated-through-hole* technique, where a hole drilled through the interconnection pads is subsequently copper-plated. The multi-layer p.c.b. production is relatively expensive and tends to be restricted to specialist highly complex circuits and systems, e.g., airborne computer systems. However, the application of *double-sided* p.c.b.s has increased considerably during the last decade, coincident with the development and application of microprocessor and microcomputer systems, so that the problems associated with the complexity of interconnections are largely overcome by this technique.

10.5 Mounting and connections

The most common method of mounting components into circuit boards is the *insertion* method, where the components have their leads passed through plated or unplated holes in the p.c.b., and then soldered, as shown in Fig. 10.3(*a*). Components having round axial leads, round radial leads and flat leads (i.e., most DIL packages) can be easily mounted by this method.

The reliability of an electronic system depends largely on the number of electrical connections involved, the most reliable system being one with the least number of connections. Permanent methods for connection are inherently more reliable than temporary 'disconnect' methods—which invariably introduce another joint in the interface area. Therefore, the designer must attempt to strike the correct balance between reliability, ease of production and servicing.

The most widely used method of making permanent connections is by *soldering*, which is the technique of joining metal-to-metal by means of a solder

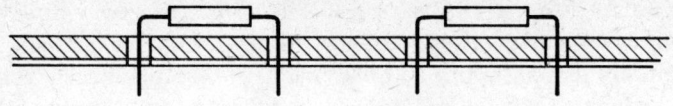

(*a*) Insertion mounting

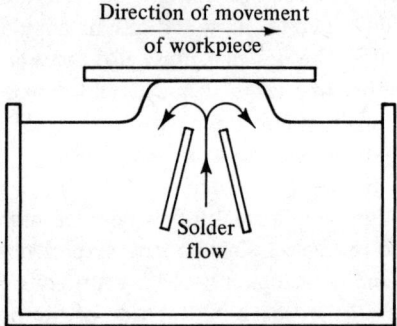

(*b*) Flow soldering

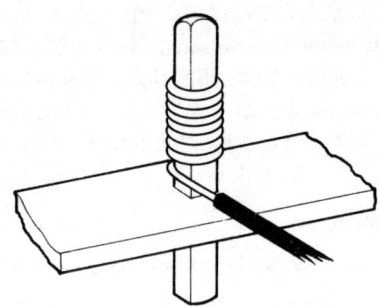

(*c*) Wire wrap connection

Fig. 10.3. Mounting and connections.

alloy employing heat and flux. The flux is used to aid heat conduction from the heat source to the workpiece, and to remove surface oxides at the instant when the molten solder 'wets' (goes into solution) the surfaces to be joined.

Electrical soldering irons are generally used for small-scale production, servicing, repair and prototype development. Large-scale production tends to favour *flow, or wave soldering*, in which boards are passed over a wave of continually circulating molten solder, generated by a special nozzle, as shown in Fig. 10.3(*b*). Ideally, the wave should have a flat crest and turbulent sides so that excess solder is removed.

A popular method of interconnection, for hard-wiring and prototype circuit boards, is the wire-wrap connection, since it provides a good quality electrical *and* mechanical connection. Single-strand wire is wrapped around sharp-edged posts under tension, as shown in Fig. 10.3(*c*). A minimum of five turns of wire is usually recommended, and can be accomplished using manual or fully automatic electric or pneumatic tools or machines.

10.6 Fault diagnosis

The broad-based techniques of fault diagnosis may equally be applied to digital electronic circuits and systems. Most electronic equipments and systems can be considered to be made up of a number of individual functional stages. The general technique of fault finding is first to locate the faulty/defective *stage*. Two methods are widely used, the *end-to-end* method and the *half-split* method.

In the *end-to-end* method, power is applied to the system and, if necessary, a signal is injected at the input. Measurements are then made at the output of each stage, working either from the output towards the input, or from input to output, until the faulty stage is located. This method is widely used on systems which do not have very many stages.

With the *half-split* method, the system under test is divided in half and each half separately tested. Assume that a system has *six* stages which are divided into two sets of three stages. Power is applied and, if necessary, a signal is injected at the input to the first stage. The output is then measured at the output of the third stage. If it is assumed that this measurement is correct, then the test is repeated on the second set of three stages, i.e., the input is applied to the input of the fourth stage while the output is measured from the output of the sixth stage. This will not produce the required result (assuming that there *was* a fault in the system). The second half of the system therefore contains the fault—so this half is further divided in half. In this case, there are three stages so the half-way point may be taken between the fourth and fifth stages or between the fifth and sixth stages. If the measurement of the output of the fourth stage is correct, but the output from the fifth stage is not correct, then the fault is in the fifth stage.

Having located the faulty stage, it now remains to determine the faulty component within that stage. This requires knowledge of the circuit operation and the function of the individual components. The process of fault diagnosis on digital circuits is often referred to as *debugging*, i.e., getting rid of the faults (or *bugs*) in the circuit. The area in which the fault lies may be located using the end-to-end and half-split methods, and the component fault location is then determined by circuit testing using a range of special test equipment.

10.7 Digital test equipment

A wide range of digital test equipment is currently available to meet the various requirements of testing and fault diagnosis on digital circuits and systems. This range extends from logic probes and pulsers, through logic clips/checkers and logic analysers to complete automatic test equipment (ATE).

Logic probes are used to detect and display logic levels by illuminating LEDs, without imposing any significant electrical load on the circuit under test. Many logic probes also incorporate the facility to detect and display pulses and voltage transients. Since different logic families have different logic levels, it is important that the correct probe is used for the correct family—this may be achieved either by choosing the most suitable probe for the family being tested, or, by choosing a probe which incorporates the facility to select the logic family being tested by an integral switch on the probe. The technique used in several popular logic probes is that the probe tip is connected to a dual *window-comparator* and an *edge-detector*. The window-comparator bias network determines the threshold levels of the logic signals—which for the DTL/TTL mode are 2.25 V (logic 1) and 0.8 V (logic 0) and for the CMOS/HTL mode are 70 per cent of supply voltage (logic 1) and 30 per cent of supply voltage (logic 0). The edge detector generally responds to both positive and negative transitions and also drives a pulse-stretcher circuit which converts level transitions and short duration pulses (typically down to 30 ns) to a steady pulse—to illuminate an LED at a frequency of, say, 10 Hz.

Logic pulsers are often packaged in a similar way to probes, and generally operate from 5 V to 15 V d.c. supply. The tri-state output, normally of high impedance, allows the user to connect the pulser tip to the node or logic gate under test without the need to desolder or cut p.c.b. track. When the pulse button is activated, a narrow pulse—swinging first logic HIGH and then logic LOW, finally returning to the high impedance state—is delivered into the node from the output of the pulser. The pulse applied changes the logic state of the node, and the resultant changes at other points in the logic system may be checked with the logic probe. Provision may be made to apply a single pulse, several pulses or continuous pulses, and the output is generally designed to sink or source up to about 500 mA.

Logic clips/checkers are devices which incorporate LEDs for indicators and actually clip onto digital ICs to give immediate indication of the logic state of each pin. The checker derives its power from the IC power supply irrespective of which pins are used, and are most useful for analysing static or low-frequency dynamic systems. Higher-frequency dynamic systems may be checked by reducing their clock rate.

Logic analysers are usually sophisticated, relatively expensive (several thousand pounds) pieces of test equipment, and are particularly useful for carrying out all the logical testing and analysis for microprocessor-based systems, computer systems and other program-controlled devices.

Special test equipment has been designed and manufactured specifically to meet particular types of tests. Logic analysers should really be mentioned in this group, in addition to sophisticated automatic test equipment (ATE) which may well provide the capability of dealing with a range of standard (and programmable) tests using the basic system, but ATE manufacturers may offer options—sometimes referred to as 'personality modules'—that are designed to enable special tests to be carried out. These ATE systems are relatively expensive, typically £20 000 to well over £100 000, with the personality modules ranging from a few hundred pounds to several thousand pounds. Therefore, in cases where only small-scale production is applied, the expense of full ATE cannot be justified and it is desirable to use low-cost, portable, dedicated types of test equipment. Polar Electronics Ltd, PO Box 97, St Sampsons, Guernsey, CI, manufacture three particularly useful audio and visual instruments: a short-circuit locator—the *Toneohm 550*; a current tracer—the *Toneohm 580*; and a faults locator—the *Toneohm 700*.

The Toneohm 550 Short-Circuit Locator uses the principle of a high-resolution milliohmmeter. The resolution (typically 2 mΩ) represents a typical value of resistance for 2 mm length of p.c.b. track. This is achieved by a Kelvin connection for a pair of hand-held probes having very sharp and hard needle points so that they can penetrate flux deposits and solder resist. Attractive features of the milliohmmeter technique are that the test is carried out on the board with no power supply connected, and the open-circuit probe tip voltage is low (typically less than 100 mV) so that no damage will be caused to any components mounted on the board. Additionally, an audio output is available, the frequency of which increases as the resistance measurement decreases, thus facilitating testing without the need to observe the digital display. This instrument may also be used as a digital ohmmeter, for resistance values up to 20 kΩ.

The Toneohm 580 Current Tracer, as shown in Fig. 10.4, is used to trace the path of current around p.c.b. track. This instrument is used on an unpowered

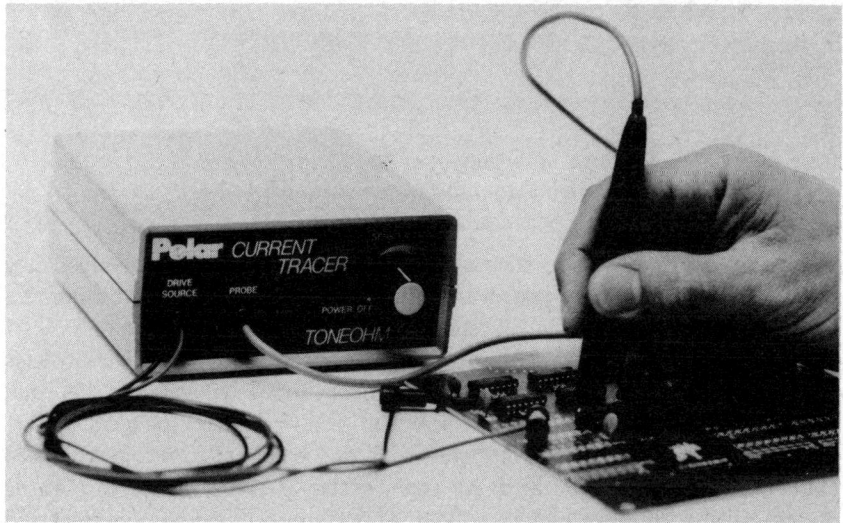

Fig. 10.4 The Toneohm 580 Current Tracer.
(with permission of Polar Electronics Ltd.)

board and its own internal low voltage a.c. drive source is connected across the two lines that are linked by a partial short, so that most of the current flows into the fault. A hand-held probe, with a magnetic pick-up coil, is placed near the tracks—the output from the detector is amplified to give an audio output signal. A tone is thus produced when the probe is in the influence of the magnetic field, so that the probe can be moved around the board to follow the current path, thus leading to the identification of the fault.

The Toneohm 700 Faults Locator, as shown in Fig. 10.5, combines the features of the Short-Circuit Locator and the Current Tracer. In addition, this instrument incorporates a microvoltmeter, that is suitable for identifying faults

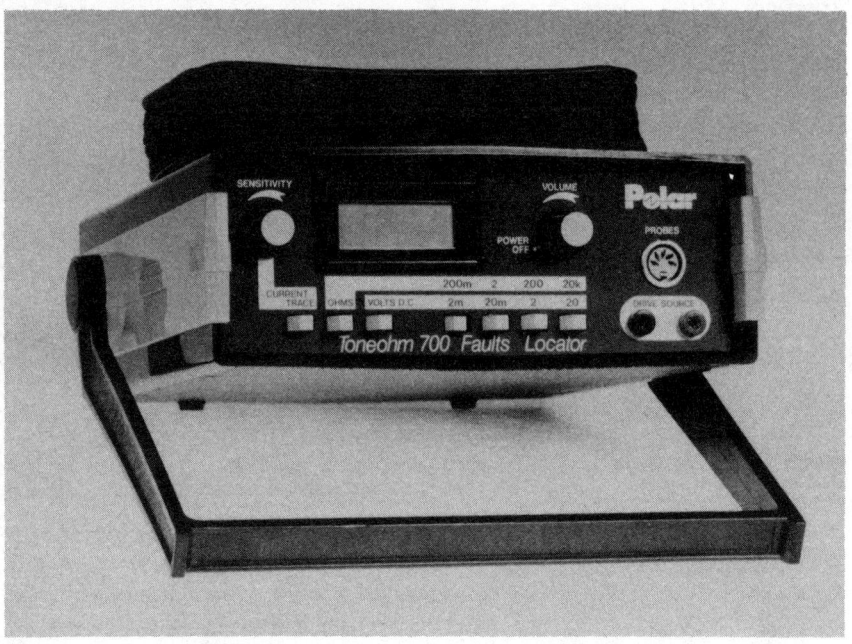

Fig. 10.5 The Toneohm 700 Faults Locator.
(with permission of Polar Electronics Ltd.)

that cause loading on power supply lines, e.g., a low-resistance fault which may disappear when the faulty system is switched off. This type of fault requires a d.c. supply for testing, either by using the faulty unit's own power supply, or, if the fault remains on the board when it is unpowered, then, by using a low-voltage d.c. drive source—derived from the instrument—connected between V_{CC} and GND to drive current around the loop. The resulting p.c.b. voltage drops are then measured using the microvoltmeter, enabling the current path to be followed until the faulty device is located. An audio output is also available, eliminating the necessity to observe meter readings. This instrument can also be used as a d.c. digital voltmeter, up to 20 V.

10.8 Typical faults

The use of high packing density circuit boards, narrow track widths, automatic insertion and wave soldering techniques increases the possibility of occurrence of faults and increases the problems associated with diagnosing the faults. Current testing methods on p.c.b.s are likely to include visual inspection, systematically unsoldering devices until the fault disappears, and may involve cutting the track on the p.c.b.—and subsequent bridging the cut track—until the fault is located. Excessive 'rework' in the form of lifted pads and cut bridged tracks may result in the p.c.b. being scrapped. It is therefore desirable to reduce the amount of rework to a minimum, which necessitates alternative testing methods to the conventional methods indicated above.

It is estimated that 50 per cent of all manufacturing defects are due to *short circuits*. P.c.b. shorts are most commonly caused by the following:

(a) solder bridges,
(b) incomplete etching during p.c.b. manufacture,
(c) component leads bent so that they touch adjacent p.c.b. track or pads, or bare leads.

Visual inspection is unlikely to reveal 'solder bridges' which are *hairline*—typically 0.02–0.03 mm wide—and may be hidden under solder resist or even hidden under IC devices or IC sockets. Furthermore, shorted adjacent IC pins may not be identified by visual inspection on p.c.b.s having a high component packing density. The high-resolution milliohmmeter is particularly useful for locating solder bridges.

The widespread use of bus-oriented electronic systems, i.e., microcomputer-based systems, has increased the occurrence of *partial device shorts* (up to 200 Ω) between tracks; *stuck node* faults in which a bus line is stuck low (or high); and *hard-shorts on multilayer p.c.b.s*, which cannot be accurately located using the milliohmmeter technique, so that an alternative principle must be used to locate accurately a fault in the middle of the p.c.b. or if the fault occurs underneath a device. Stuck node and partial device shorts aggravate the problem of identifying the actual IC which is faulty. When the ICs are mounted in IC sockets, it is relatively easy to remove one at a time until the faulty one is identified, but if the ICs are soldered into the p.c.b., it is much more time-consuming, and damaging, to remove ICs. The arguments for and against the use of IC sockets are likely to be debated indefinitely by the various factions—two of the main considerations being cost and reliability. The current tracing principle is therefore extremely useful as an aid to locating faulty components.

Loaded supply (V_{CC}) line faults frequently occur, in which the supply line is loaded, pulling it down to a low level. This may often be caused by a low resistance fault, which may disappear when the unit under test is switched off. Current tracing methods are not really suitable for this type of fault—mainly due to the low reactance of decoupling capacitors causing false current paths. There-fore, a d.c. test must be used, either by using the unit's own power supply, or, if the fault remains when the board is unpowered, by using a low-voltage d.c. drive connected between V_{CC} and GND to drive current around the fault loop whilst a microvoltmeter is used to measure the p.c.b. voltage drops.

10.9 Techniques of fault finding

Unfortunately, there is no test equipment available which will act as the 'magic wand' of fault diagnosis, i.e., no equipment can identify the faulty stage, or board, *and* then identify the location of the fault on that board. It must, therefore, be *emphasized* that fault diagnosis is an area in which the technician and engineer must attempt to acquire expertise in the *application* of the available diagnostic instrumentation. Sophisticated ATE equipment can be used to identify the faulty board in a complex system, and even identify the type of fault which exists on that board, *but*, without any add-on 'personality' modules, this expensive equipment will not be able to tell the user the actual location of the fault on the board.

Therefore, once a faulty board has been identified, a relatively low-cost technique of locating the actual fault is advantageous. If the type of fault has also been identified, then this does make the fault location tests somewhat less arduous. The onerous tasks of fault diagnosis can be lightened by applying logical reasoning to the sequences of tests *and* to the results of the tests, since it is generally the analysis of the tests, rather than the results themselves, that reveals the location of the fault.

10.10 Short-circuit faults

If it is supposed that a board has been identified as having a short-circuit (from the application of ATE) or, a board is suspected as having a short-circuit, then the *Toneohm 550 Short-Circuit Locator*, or the *Toneohm 700 Faults Locator* may be used.

Assuming that the *Toneohm 700 Faults Locator* is to be used, then proceed as follows:

(a) Select OHMS and the 200 mΩ range.
(b) Connect the Kelvin probes into the PROBES socket.
(c) Remove the power supplies from the faulty p.c.b.
(d) Probe the suspected tracks; when two shorted tracks are probed, a tone will be audible.
   Note It may be necessary to adjust the VOLUME before satisfactory results are obtained. A useful technique is to probe two points (about 2–3 cm apart) on the same track to make the initial adjustments.
(e) Move one probe along a track and note the change in the frequency of the tone (or the value of the display). A higher frequency (or lower display reading) indicates that the probe has been moved closer to the short fault.
(f) Continue moving one (or both) of the probes until the highest frequency (or lowest reading) is obtained, as shown in Fig. 10.6, in which the probes were moved as follows:

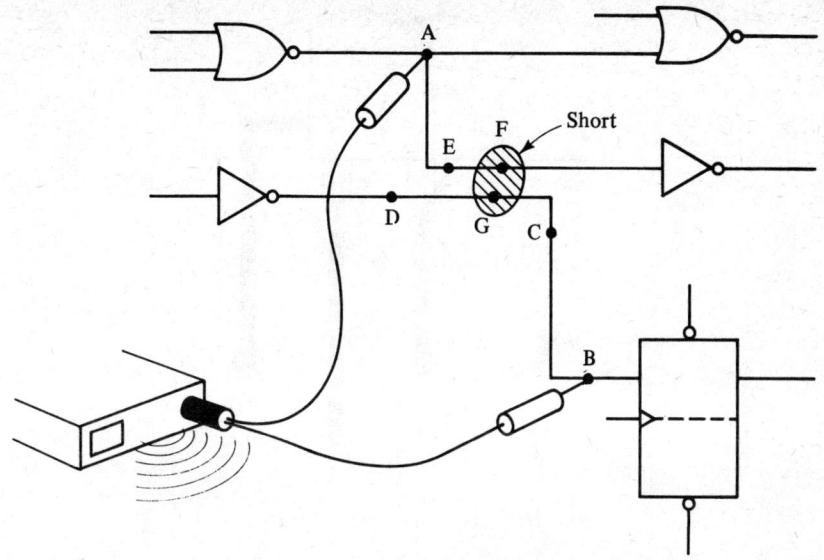

Fig. 10.6. Location of short-circuit.

    (i) probes at points A and B—a tone is audible;
   (ii) probes moved to points A and C—the pitch of the tone increases, i.e., closer to the fault;
  (iii) probes moved to points A and D—the pitch of the tone decreases, i.e., further from the fault;
  (iv) probes moved to points E and C—the pitch of the tone increases—closer to the fault;
   (v) probes moved to points F and C—the pitch of the tone increases—closer to the fault;
  (vi) finally—in this case—the probes are moved to points F and G—and the pitch of the tone is at its highest (and display shows the lowest reading).
(g) When the pitch of the tone is at its highest (or meter displaying its lowest reading), the positions of the probes are within 2–3 mm of the short faults.
(h) Finally, the fault must be cleared. This may be achieved by cutting between the tracks, if the fault was a hairline solder bridge, as shown in Fig. 10.7.

10.11 Partial device shorts and stuck nodes

If it is assumed that a board has been diagnosed as having a partial device short, i.e., of the order of 1–200 Ω—which may not be immediately obvious—the symptoms may be similar to those of a short. If the fault is a 'hard-short' then the milliohmmeter technique described in Sec. 10.10 may be used; however, the fault may often be a low-resistance fault which may disappear when the unit under test is switched off.

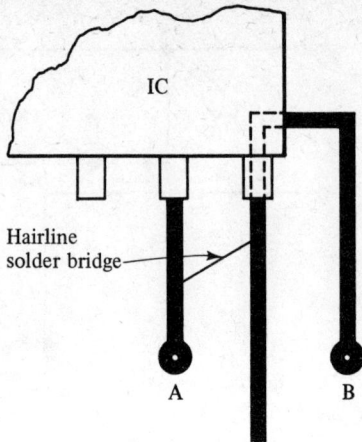

Fig. 10.7. Hairline solder bridge.

Two techniques are recommended for location of this type of fault, the current tracing and microvoltmeter methods. It is useful to be able to perform a 'quick test' to determine the general area of the fault, before proceeding to the detailed testing procedures. It is therefore recommended that the current tracing technique is applied first, followed by the microvoltmeter technique. The first method may use the *Toneohm 580 Current Tracer*; or the *Toneohm 700 Faults Locator* may be used as follows:

(a) Remove the power supplies from the faulty p.c.b.
(b) Select CURRENT TRACE, connect the low voltage (0.55 V) DRIVE SOURCE leads across the suspect pair of tracks which have been loaded down by the partial fault, and connect the current tracer probe into PROBES.
(c) Place the current tracer probe near the faulty bus track and adjust the SENSITIVITY and VOLUME until a tone is audible.
    Note For a quick-scan test, hold the probe about 1 cm from the track and adjust to give an audible tone.
(d) Move the probe away from the track, and the tone becomes inaudible. SENSITIVITY may be adjusted for optimum sensing distance of the probe from the p.c.b. track.
(e) Move the probe along the faulty track, using the tone for guidance, i.e., if the tone becomes inaudible, the probe has moved away from the faulty track.
(f) Eventually, the current will be found to enter a faulty device on the bus.

Occasionally, when current tracing has been used to trace a bus fault, the particular p.c.b. track layout may prevent exact identification of the faulty device, and only isolate it down to two or three possible ICs. It is then necessary to use the microvoltmeter technique to pinpoint exactly which IC is sinking the current.

Furthermore, for the low-resistance fault which disappears at switch off, the current tracing technique is not really suitable, since the low reactance of de-coupling capacitors can cause many false paths for the a.c. drive source current to flow through. Thus, tests for this type of fault must be performed at d.c., either by using the equipment's own power supply current or, if the fault remains with the faulty p.c.b. unpowered, then a low voltage d.c. drive source connected between V_{CC} and GND may be used to drive current around the faulty loop. This type of test may be performed using the *Toneohm 700 Faults Locator* as follows:

(g) Assuming that the p.c.b. under test is unpowered. Connect the low voltage (0.55 V) d.c. DRIVE SOURCE leads across the V_{CC} and GND of the suspect board, and connect the Kelvin probes into the PROBES socket. Select VOLTS and range 2 mV (or 20 mV).

(h) This type of test is useful for making relative measurements rather than absolute measurements, since p.c.b. track thicknesses will affect the actual voltage drop. Therefore, the use of the tone simplifies the use of the tester. The tone is proportional to the measured voltage drop on these two ranges.

(i) Assume that a board contained a faulty tantalum capacitor on a V_{CC} rail as shown in Fig. 10.8.

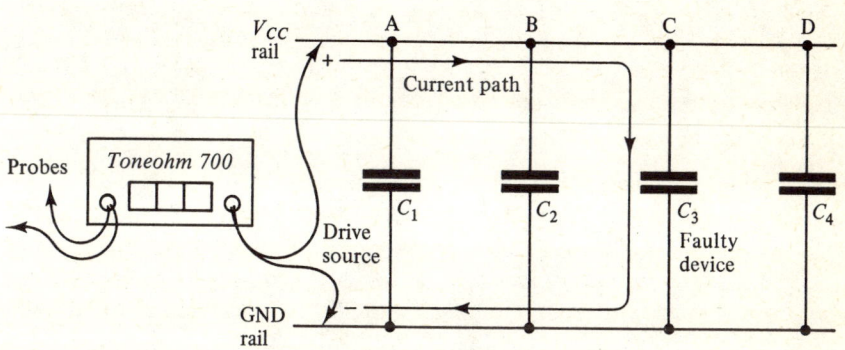

Fig. 10.8. Isolation of partial V_{CC} fault.

(j) The resistance measurement in this case is 8 Ω, between V_{CC} and GND, so that the external d.c. drive source is connected as shown in Fig. 10.8, and with the *Toneohm 700* set up as described above, successive probe measurements are made between V_{CC} and nodes A, B, C and D.

(k) The voltage measurements obtained are as follows:

A   0.205 mV
B   0.427 mV
C   0.785 mV
D   0.785 mV

(l) The last of these measurements indicates that there is no voltage drop in the track between nodes C and D. A further measurement between C and D confirms this.

(m) Therefore the faulty capacitor is located as C_3.

10.12 Summary

The location of the faults beyond the node affects the complete electronics industry from original manufacture of the p.c.b. through to customer returns from field service. Although complex and sophisticated ATE can diagnose these faults efficiently, the faults described can only be accurately located with a minimum of rework and expense by using stand-alone equipment similar to that described above.

Proficiency at fault diagnosis can only be achieved by a great deal of practice. It is recommended that as much experience as possible is gained on fault diagnosis, using as wide a range as possible of test equipment.

Appendices

A Comparison of standard logic symbols

A selection of symbols which have been widely used for more than a decade are shown in Fig. A.1. The most useful are those used by the US Military Standard, since this is the standard that most manufacturers work to. Hence, because of the necessity to refer to manufacturer's data, we must be aware of the symbols they use.

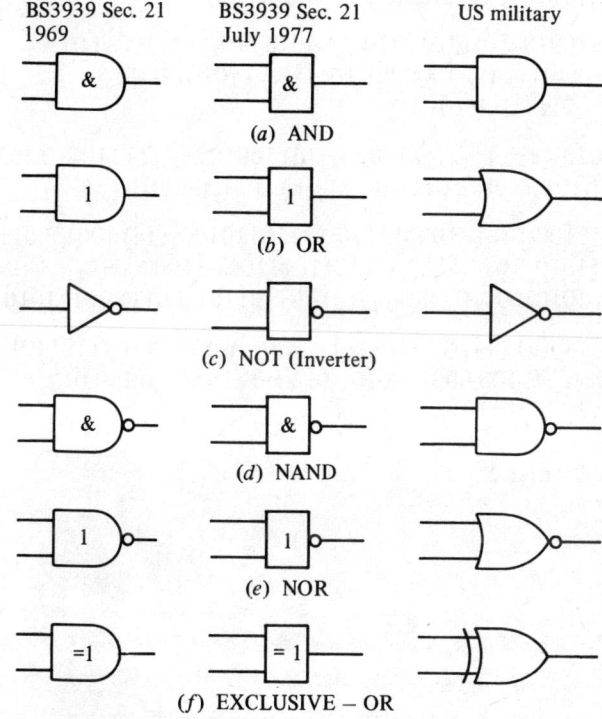

Fig. A.1. Comparison of standard logic symbols.

B Solutions to numerical exercises

Exercises to Chapter 1

1. (a) 10001_2, (b) 100011_2, (c) 100101.0001_2, (d) 11100.101_2,
   (e) 11011101.11_2.

2. (a) 13_8, (b) 27_8, (c) 34.7_8, (d) 205.03_8, (e) 373.14_8.

3. (a) 9H, (b) EH, (c) 49.EH, (d) FF.AH, (e) 84.2H.

4. (a) 11_{10}, (b) 27_{10}, (c) 6.375_{10}, (d) 23.5625_{10}, (e) 215.21875_{10}.

5. (a) 17_{10}, (b) 31_{10}, (c) 3.953125_{10}, (d) 91.453125_{10}, (e) 189.15625_{10}.

6. (a) 9_{10}, (b) 235_{10}, (c) 79.34375_{10}, (d) 243.921875_{10}, (e) 164.734375_{10}.

7. (a) 11101_2, (b) 111110011_2, (c) 110101.11101101_2,
   (d) 11101111.0100011_2, (e) 100111111.110101011_2.

8. (a) 10111_2, (b) 11000011_2, (c) 111110.0111101_2,
   (d) 1110110.10101111_2, (e) 1111.111001_2.

9. (a) 5_8, (b) 23_8, (c) 27.5_8, (d) 12.26_8, (e) 335.54_8.

10. (a) 9H, (b) 15H, (c) 36.DH, (d) 57.D8H, (e) F7.EH.

11. (a) 11110010_2, (b) 1011010_2, (c) 101110_2, (d) 100111000_2,
    (e) 1111.101_2, (f) 1000001.1011_2, (g) 10101.0011_2,
    (h) 11100111.01110101_2.

12. (a) 10010_2, (b) 11_2, (c) 10010_2, (d) 1001011_2, (e) 110110.10101_2,
    (f) 111.101_2, (g) 1010.001011_2, (h) 1011000.1101_2.

13. (a) 1010100_2, (b) 1100011_2, (c) 1110110100_2, (d) 1001100111000110_2,
    (e) 101101011011.011_2, (f) 10110111001.110000001_2,
    (g) 1101010100000.00011_2, (h) $1011011011001111.01110101111_2$.

14. (a) 1001_2, (b) 11_2, (c) 110_2, (d) 1000_2, (e) 110001.11101101_2 to eight
    places, (f) 100.000100010_2 to eight places, (g) 100.01_2,
    (h) 1111.00010110_2 to eight places.

Exercises to Chapter 2

1. $A + B$

2. $A + B$

3. 0

4. $A.B$

5. $A + B$

6. 0

7. 1

10. $A. (B + C)$

11. $A.\bar{B}.C$

12. $A.\bar{C}$

C Pin-outs of a selected range of 74 series TTL ICs

The pin-out connections are shown in Fig. C.1 for a *selection* of 74 series TTL ICs. The connections are always labelled as viewed from the top of the package. Always check with manufacturers for complete data, and for a complete list.

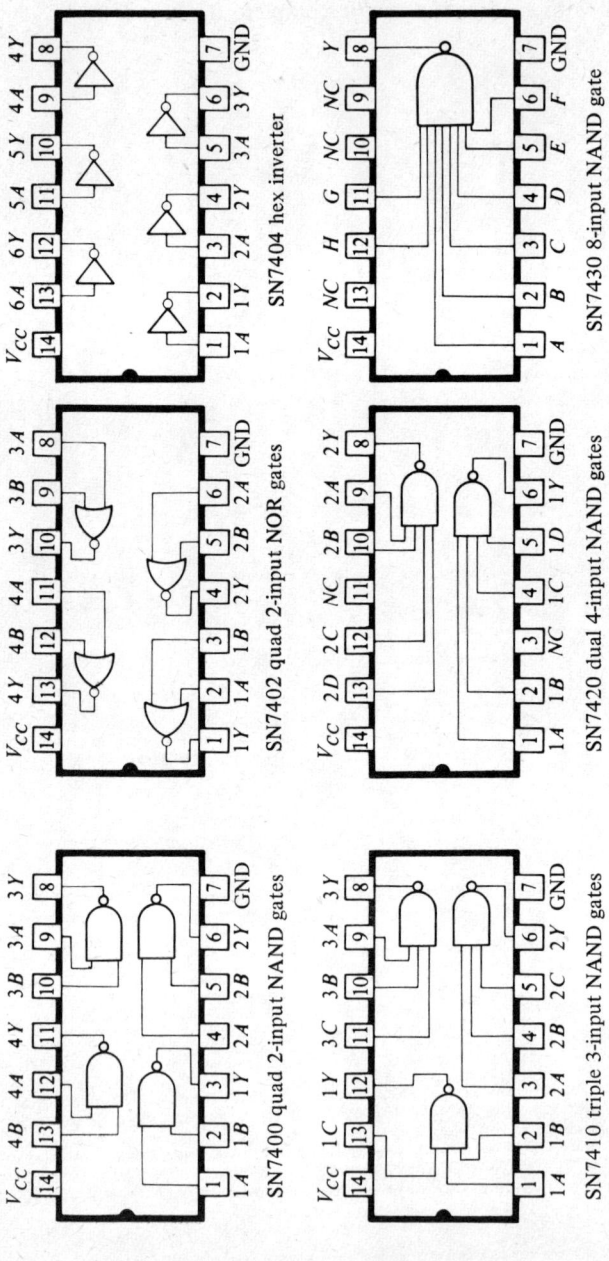

Fig. C.1. Pin-out connections for 74 Series TTL ICs.

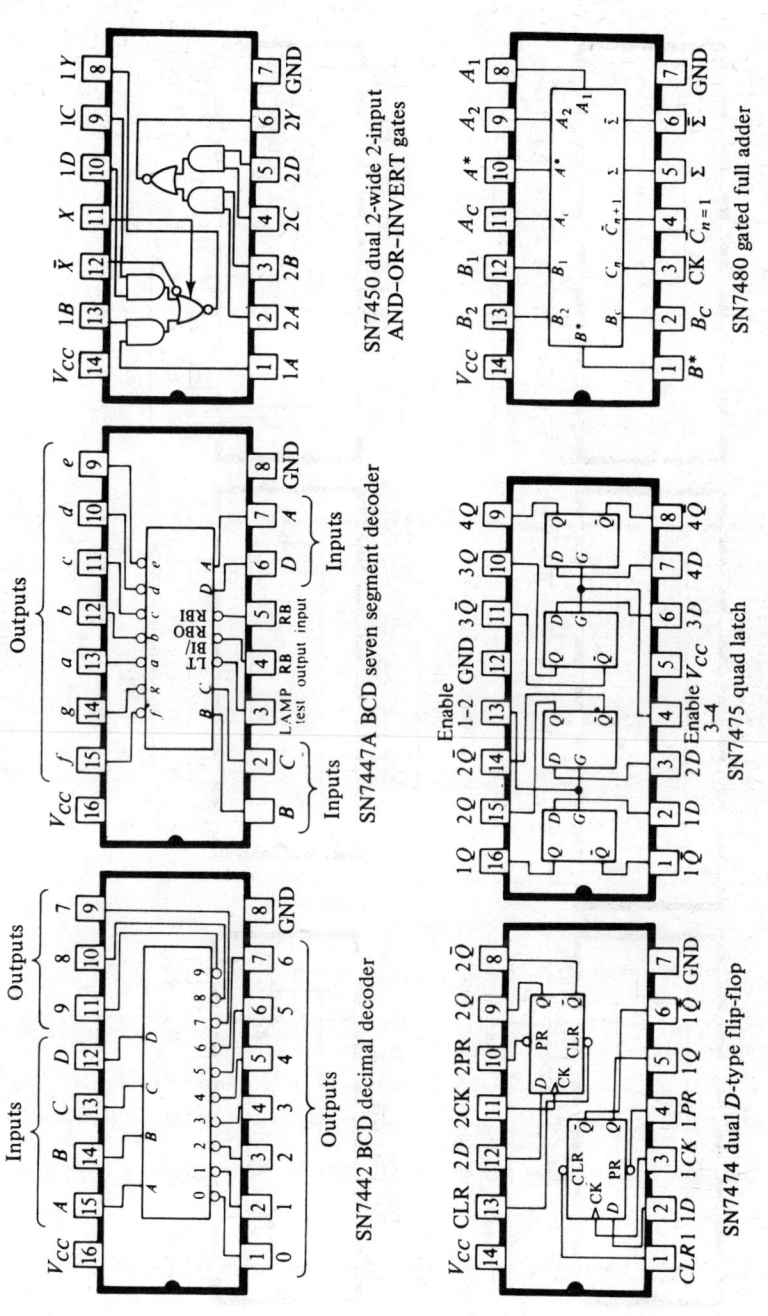

SN7450 dual 2-wide 2-input AND–OR–INVERT gates

SN7480 gated full adder

SN7447A BCD seven segment decoder

SN7475 quad latch

SN7442 BCD decimal decoder

SN7474 dual D-type flip-flop

Fig. C. 1. continued.

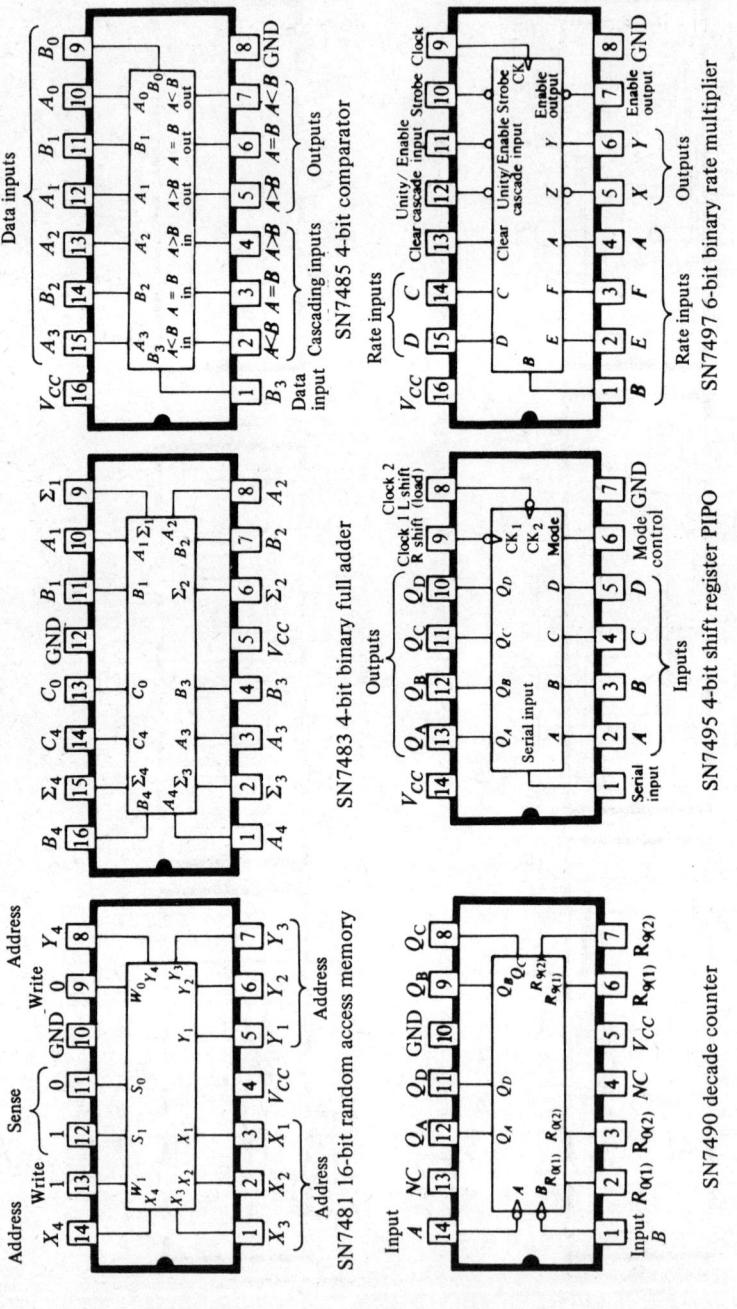

Fig. C.1. continued.

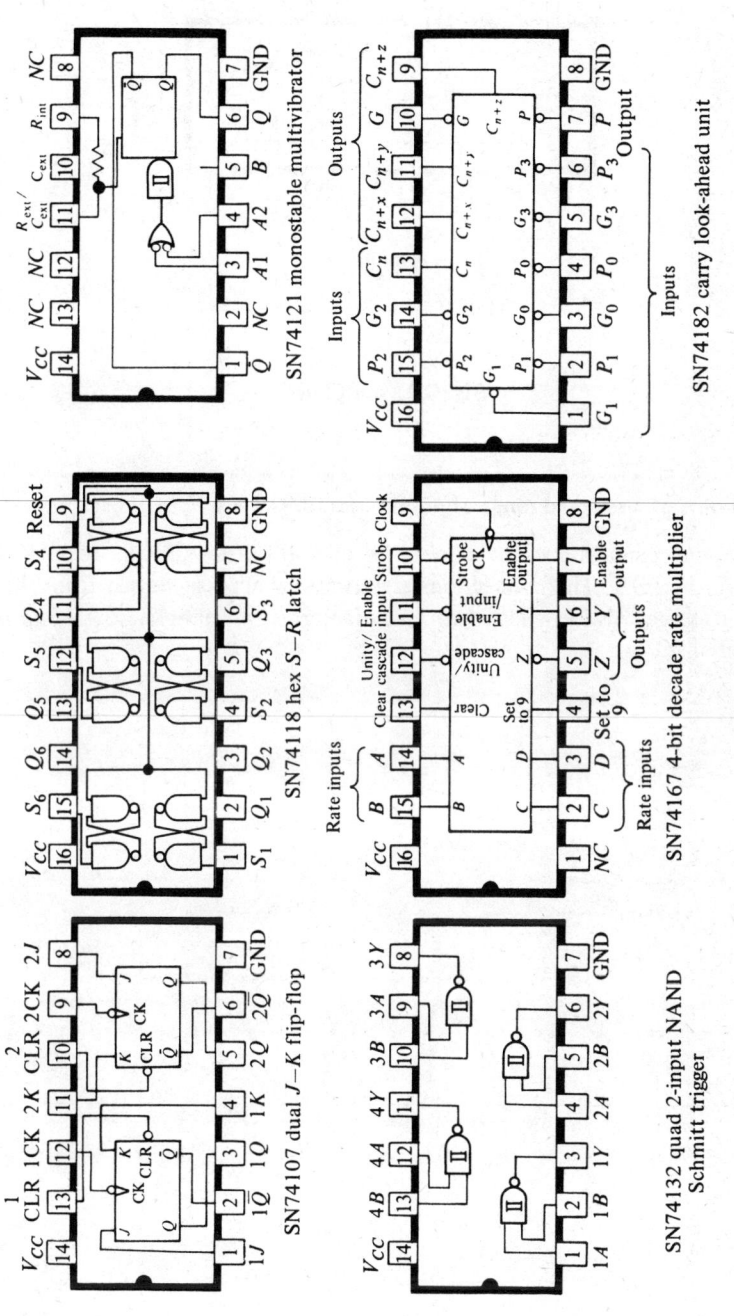

Fig. C.1. continued.

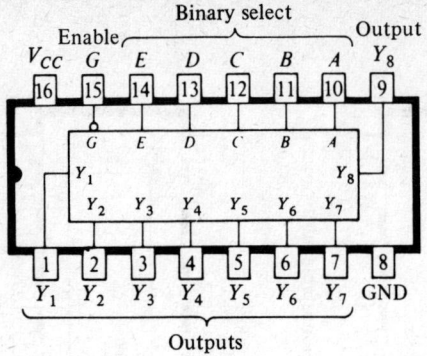

SN74188 256-bit PROM fusible

Fig. C.1. continued.

D   Pin-outs of a selected range of 4000 series CMOS ICs

The pin-out connections for a *selection* of 4000 series CMOS digital ICs are shown in Fig. D.1. The connections are always labelled as viewed from the top of the package. Always check with manufacturers for complete data, and for a complete list.

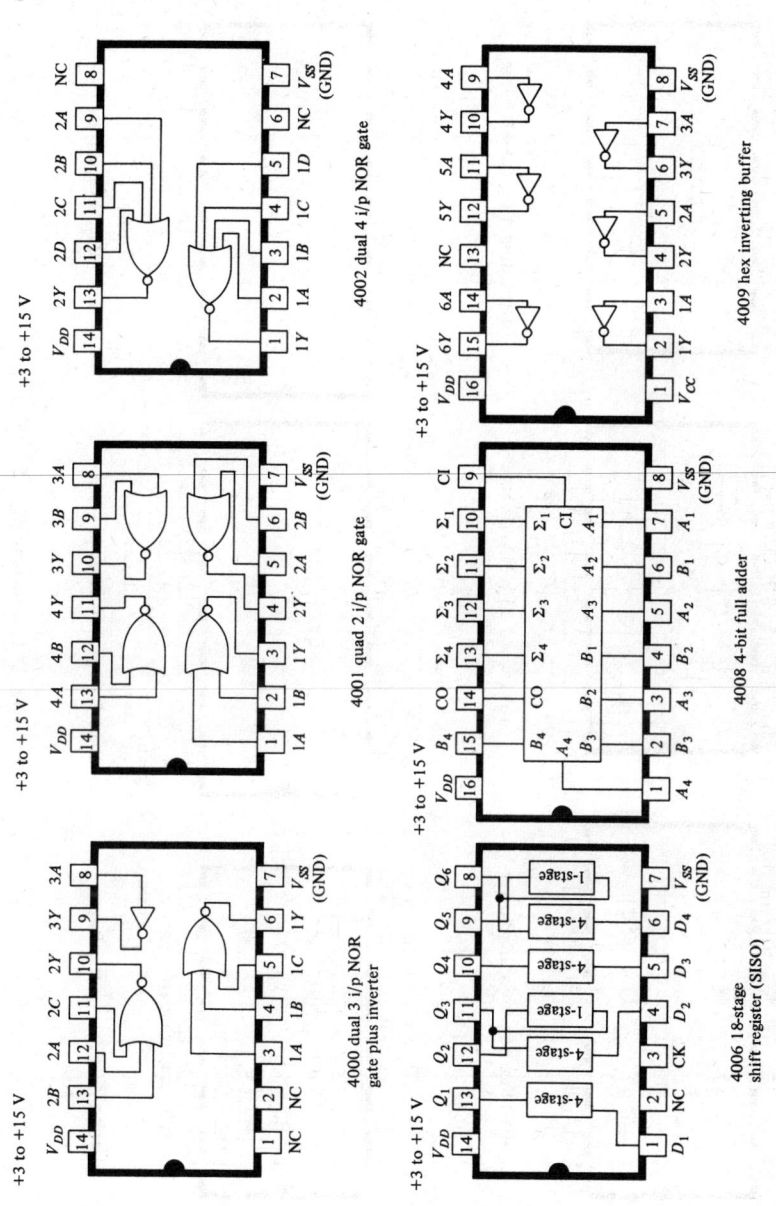

Fig. D.1. Pin-out connections for 4000 series CMOS ICs.

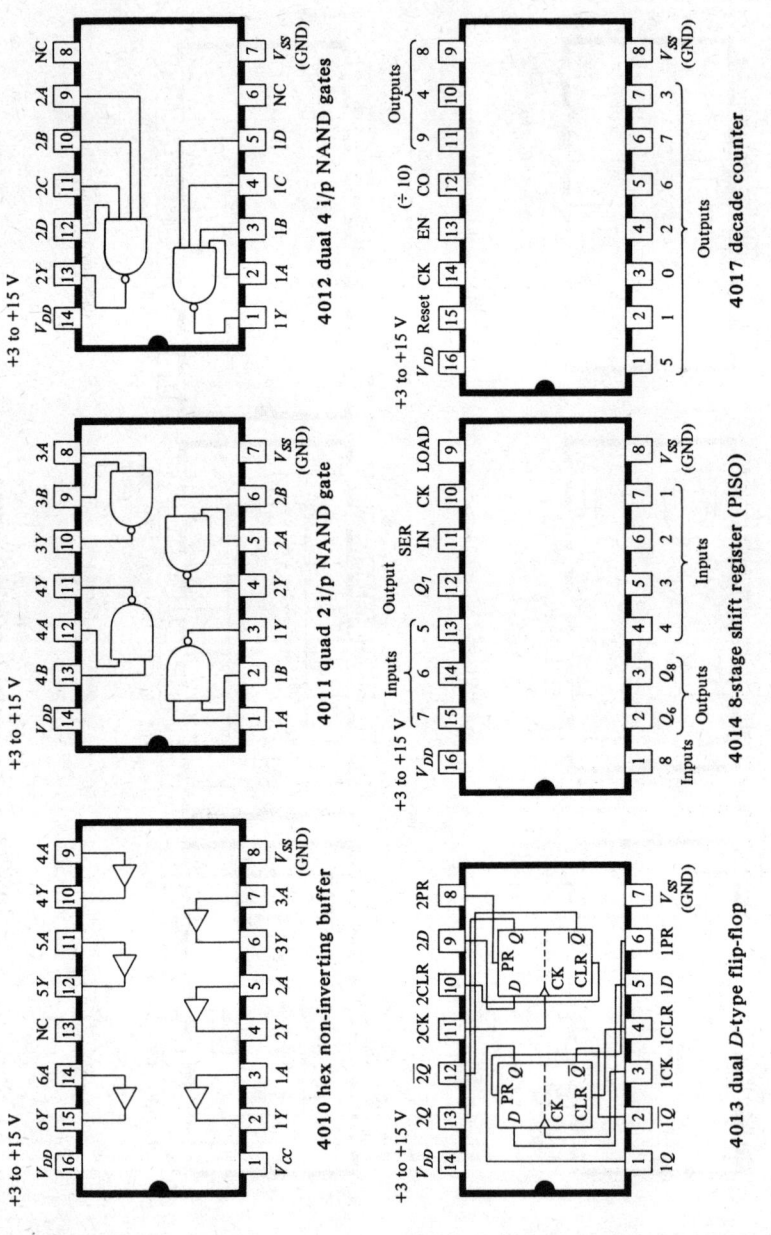

Fig. D.1. continued.

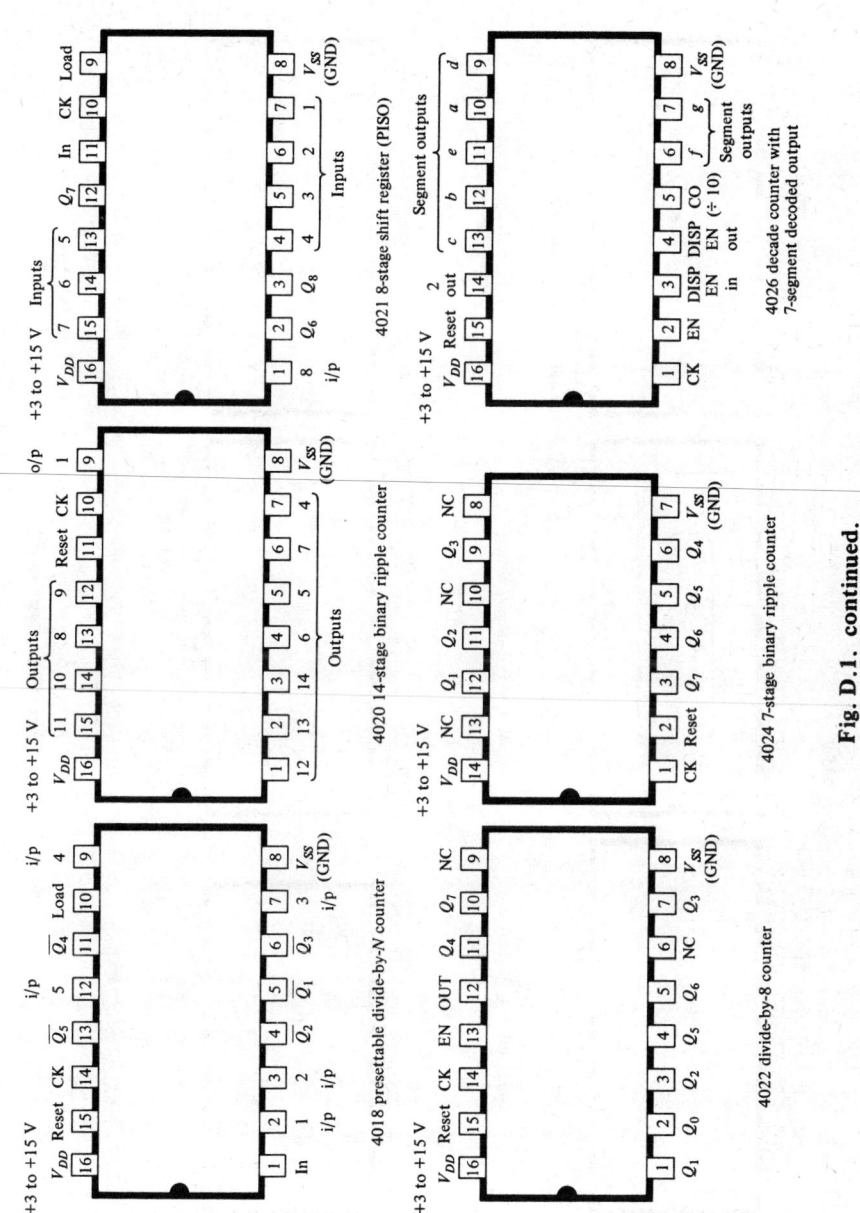

Fig. D.1. continued.

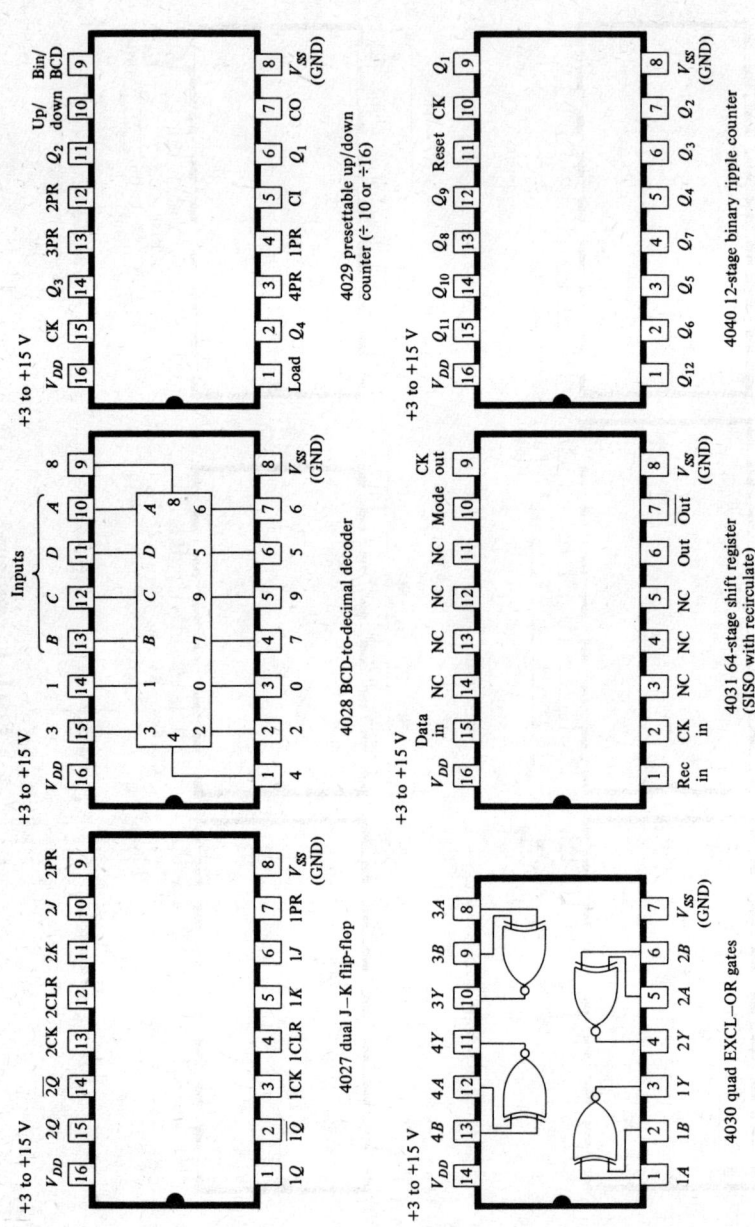

Fig. D.1. continued.

Index